THE ATLANTIC COAST

Harry Thurston

Photography by Wayne Barrett

Illustrations by Emily S. Damstra

# THE ATLANTIC COAST

## A NATURAL HISTORY

David Suzuki Foundation

GREYSTONE BOOKS
D&M PUBLISHERS INC.
*Vancouver/Toronto/Berkeley*

*Advisory Panel*

First U.S. edition 2012

11 12 13 14 15 5 4 3 2 1



Greystone Books
An imprint of D&M Publishers Inc.
2323 Quebec Street, Suite 201
Vancouver BC Canada V5T 4S7
www.greystonebooks.com

David Suzuki Foundation
219–2211 West 4th Avenue
Vancouver BC Canada V6K 4S2

Cataloguing data available from Library and Archives Canada
ISBN 978-1-55365-446-9 (cloth)
ISBN 978-1-55365-965-5 (ebook)

Editing by Nancy Flight
Copyediting by Peter Norman
Jacket and interior design by Naomi MacDougall
Front and back jacket photographs by Wayne Barrett
Maps by Eric Leinberger
Printed and bound in China by C&C Offset Printing Co., Ltd.
Text printed on acid-free paper
Distributed in the U.S. by Publishers Group West

We gratefully acknowledge the financial support of the Canada Council for the Arts, the British Columbia Arts Council, the Province of British Columbia through the Book Publishing Tax Credit, and the Government of Canada through the Canada Book Fund for our publishing activities.

# PREFACE

WHEN I WAS a child, my family often took a Sunday drive from our home in Yarmouth, Nova Scotia, to Cape Forchu to visit the light keeper. This "forked cape," named by Samuel de Champlain when he sailed by it in 1604, points into the Gulf of Maine and toward the eastern seaboard of the United States, or the "Boston States," as Nova Scotians have traditionally called New England. Once you pass the cape's gray rocks, heading in a northeasterly direction as Champlain did, you might properly be thought of as having entered the Bay of Fundy. Looking the other way, to the southwest, on a clear day you can see the silhouette of Green Island, rising darkly above the waters in the distance, the first in a chain of islands that guard the south coast of Nova Scotia, which faces the North Atlantic proper.

On occasion, I got to climb the dizzying staircase of the old Cape Forchu lighthouse for a panoramic view of the waters that nourished the surrounding coastal communities. Southwestern Nova Scotia was and still is home to a thriving lobster fishing industry and also home port to fleets of vessels that pursue groundfish like cod, haddock, and pollack, as well as halibut, swordfish, herring, and scallops. On a Sunday, however, these boats would be berthed in snug coves and behind breakwaters. The only boat in sight might be the Bluenose Ferry plying between Bar Harbor, Maine, and Yarmouth, Nova Scotia, its white hull suddenly appearing on the

‹ Shrouded in fog and guarded by cliffs, the Atlantic coast presents a beguiling but formidable face.

The Cape Forchu light stands guard over the rocky coast of southwest Nova Scotia.

far horizon and soon passing close by the cape, where we would wave at the passengers at the rails as other children might greet a passing train.

Most days, while the adults talked inside at the light keeper's house, I played on the cape's jagged rocks, scrambling over ridged backs that could cut like a knife, inspecting tide pools created when winter waves crashed over this thin finger of land, and clinging to the stunted, salt-sprayed white spruce that grew along the cliff edges. Below, the waves churned, marbling the water and swirling thick mats of rockweed.

This place became so close to my heart that later I chose to be married there. A thick fog shrouded the outdoor ceremony with secrecy and mystery. A crop of purple irises at our feet lit up the noon hour grayness with their blue flames, while the lighthouse beacon rotated and the foghorn sounded a rhythmic blessing.

The cape is a symbol of the central role that the Atlantic has played in my life. The ocean has never been far away as I moved from my childhood home, where the sea encroached upon and created the tidal marshes of the Chebogue River, to Acadia University, where I studied biology overlooking the

Minas Basin in the upper reaches of the Bay of Fundy, and finally to my present home on the tidal Tidnish River, near the Northumberland Strait of the Gulf of St. Lawrence.

For the last three decades, in my role as an environmental journalist and natural history writer, I have had the privilege of exploring firsthand much of the diverse and dramatic Atlantic coastline, from northern Labrador, where snow-capped mountains overlook Greenland-born icebergs in August, to the beaches of Delaware Bay, where, every May at the full moon, a brightly colored phalanx of shorebirds waits for the coming ashore of an armada of primitive horseshoe crabs—a phenomenon that predates the creation of the modern-day Atlantic itself. I have followed fishermen to a heaving offshore world as they dragged a bounty of scallops from the seabed of Georges Bank and biologists as they studied the unique flora and fauna of the remotest of the Northwest Atlantic islands—Sable, where so many ships crossing this northern ocean foundered in the days of sail. I have collected eiderdown on the seabird islands in the St. Lawrence River estuary and dulse on the shores of the Bay of Fundy at low tide. In Newfoundland, I have watched, amazed and amused, as schools of smeltlike capelin blackened the inshore waters and came ashore to mate on the beaches, while whales, cod, and seabirds congregated just offshore to gorge on this annual massing of little fish.

Despite the great fecundity of these waters, it has been impossible to ignore the depredations this marine ecosystem has been subjected to in the last half century. As a summer student working in a fisheries research laboratory in the late 1960s, I was witness to the decimation of the once-great Georges Bank herring stock, as each evening the seiner fleet left the Yarmouth harbor empty and returned loaded to the gunnels with a silver bounty—until there were no more herring to catch and grind into fish meal. In the 1990s, I was shocked, as was the world at large, when the northern cod stocks—a resource that had fed the world for half a millennium and had been the economic raison d'etre for the opening of the North American continent in the first place—collapsed from overfishing. Yet the North Atlantic remains a place of marine riches, host to great immigrations and massing of marine creatures that mark the year in the life of this ocean. Great whales return to these northern waters to fatten, as do shorebirds on their annual migrations between their Arctic breeding and southern wintering grounds. And every year millions of seabirds congregate along the coastal cliffs and islands to lay their eggs and fledge their young, and each year seabird nations—northern breeders in winter and austral breeders in summer—join in great rafts offshore.

In trying to understand the ecological workings of this ocean just beyond my doorstep, I owe a debt to a long list of naturalists and biologists, who may not be mentioned by name in the chapters that follow. They include Sherman Blakeney, Professor Emeritus of Biology at Acadia University, Graham Daborn, Founder of the Acadia Centre for Estuarine Research, and fisheries biologist and Acadia professor Michael Dadswell; Canadian Wildlife Service (CWS) seabird biologists Richard Brown, Anthony Locke, and David Nettleship; the late David Gaskin and Joseph Gerardi, marine mammal experts, both formerly of the University of Guelph; CWS shorebird biologist Peter Hicklin and naturalist Mary Majka; conservationist and biologist Jean Bédard of Université Laval; Memorial University of Newfoundland animal behaviorist Jon Lien, who dedicated untold time and energy to saving both entangled whales and fishers' gear, and his colleague, ornithologist William Montevecchi. And not least, I owe thanks to other nature writers who have shown what is possible when exploring the Atlantic through language—Farley Mowat, Harold Horwood, Evelyn Richardson, Silver Donald Cameron, and Franklin Russell, to name but a handful. A special thanks to my friend David Jones for access to his extensive personal library on matters coastal and historical. I am, of course, grateful as always to work with my longtime editor, Nancy Flight, and publisher Rob Sanders, who continue to dedicate themselves to the literature and preservation of the natural world.

*facing page*: Pounding surf and veils of spray mark the margins of the Atlantic coast.

AS A CHILD, when I played perilously on the rocks of Cape Forchu, I had no notion that the ocean before me had not always been there. Nor could I have understood that the raspy solid grit underfoot had once been the ash that poured from a chain of volcanoes when the North American and European continental plates squeezed together 430 million years ago and that this deposit of volcanic tuff extended for some 5,000 meters (15,000 feet) under me. Or that the Atlantic Ocean itself began to open some 200 million years ago, before which I might have walked on solid land from Nova Scotia to Morocco. I knew only that if I made a misstep, I would plunge into the cold green waters below.

# THE ATLANTIC REALM

## *Where North Meets South*

FOG SEEMED LIKE a constant companion when I was growing up. My earliest memories are wrapped in it as if in a comforting blanket. I grew up in Yarmouth, Nova Scotia, at the very southwestern tip of that peninsular province that points into the Gulf of Maine and is all but surrounded by water. Yarmouth averages 120 foggy days a year, and the fog produced here by the interaction of land, sea, and air is often of the "pea soup" variety. When summer air, warmed by the land, flows out over the tidally generated colder waters of the outer Bay of Fundy, its moisture condenses, producing a fog bank. This fog would often burn off by midmorning, under the heat of the rising sun, but move in again when sea breezes blew onto the land in the evening. Companion to some, curse to others, to me the fog is the sea's breath, a reminder of the closeness of the sea and its influence on all life in the Atlantic realm.

The Northwest Atlantic region of North America—the Northeastern United States, the Maritime Provinces in Canada, coastal Quebec, and Newfoundland and Labrador—belongs to the sea. The maritime influence on climate, and therefore on flora and fauna, is dominant—even far inland, out of sight and sound of the ocean, and even though weather systems generally move easterly off the continent. The prevailing wind is from the southwest in summer and from the northwest in winter, though strong cyclonic storms—the famous nor'easters that blow in off the Atlantic—occasionally blast

‹ Twin lights keep watch over the waters at Cape Ann, Massachusetts.

the coast. The continental weather systems would normally make the winters long and cold and the summers very hot, but the ocean moderates these extremes. It is slow to heat up but once warmed maintains its heat longer than the land, with consequences for the duration and intensity of the seasons. The relative warmth of the ocean causes warmer weather to linger in the autumn—the so-called Indian summer—and makes the winter less severe than it is inland. In spring, however, the ocean has the opposite effect. While the land heats up more quickly, the cooler ocean causes the spring to be delayed and summer near the coast to be cooler and shorter than it is farther inland.

This maritime influence is greater in the coastal areas most exposed to the open ocean, such as Newfoundland and Nova Scotia, than in the more protected areas like the Gulf of Maine and Gulf of St. Lawrence. The latter, however, freezes over in winter, whereas in summer the shallow coastal area around eastern New Brunswick and Prince Edward Island boasts the warmest waters north of the Carolinas. Farther north, landfast ice clings to the Labrador coast from December until at least April, and icebergs, born in Greenland, drift into Newfoundland waters late into the summer before succumbing to the warm waters of the Gulf Stream off the Tail of the Grand Banks.

*facing page*:
MAP: Major current systems of the northwest Atlantic

*Cross Currents*

The Northwest Atlantic coast extends from the northern tip of Labrador, at Cape Chidley, where a treeless tundra prevails and polar bears and walrus haunt the coast, to Cape Hatteras, where the tropically warmed waters of the Gulf Stream brush the Outer Banks of North Carolina. The predominant oceanographic influence on this vast coastal region is the Labrador Current, colorfully described by the Newfoundland artist Christopher Pratt as "a relentless flood of molten ice, the bloodstream of our near sub-Arctic climate." The Labrador Current is created when cold waters from Hudson Bay and the Davis Strait converge off Cape Chidley and flow southward along the Labrador coast. It consists of two branches—a warmer, saltier offshore branch, which forms a counterclockwise gyre in the Labrador Sea, and a fresher, colder inshore branch, which wraps Newfoundland in an icy embrace.

This inshore branch itself bifurcates, one arm turning into the Gulf of St. Lawrence and the main flow continuing southward and westward. It is joined by the great flush of water that originates in the Great Lakes and flows out of the Gulf of St. Lawrence through the Cabot Strait, where it hugs the Atlantic coast of Nova Scotia as the Nova Scotia Current. This cool, relatively freshwater mass skirts the eastern half of the coast before it abruptly branches

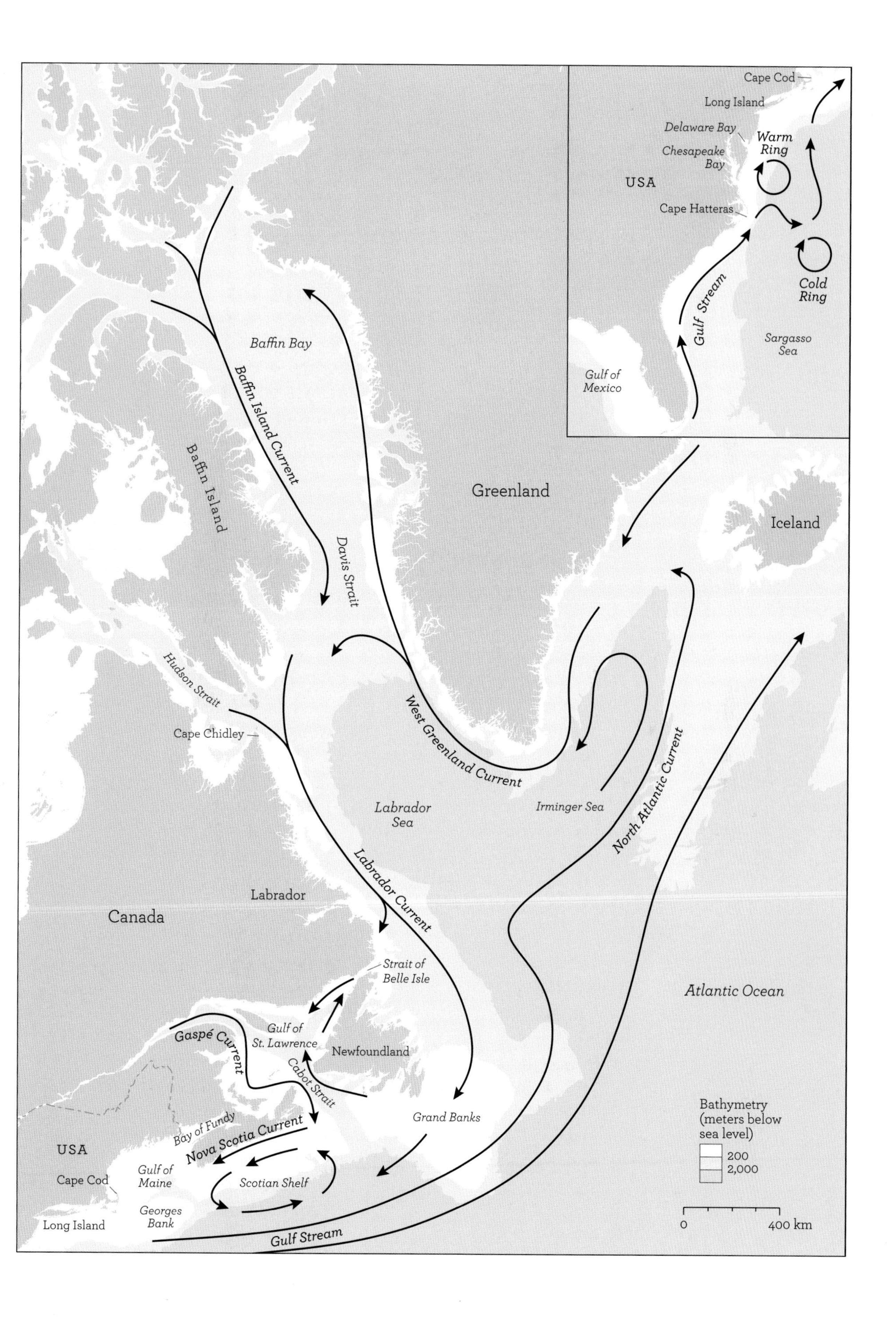
Cape Cod
Long Island
Delaware Bay
Warm Ring
Chesapeake Bay
USA
Cape Hatteras
Cold Ring
Gulf Stream
Sargasso Sea
Gulf of Mexico
Baffin Bay
Baffin Island Current
Baffin Island
Greenland
Iceland
Davis Strait
Hudson Strait
Cape Chidley
West Greenland Current
Labrador Sea
Irminger Sea
North Atlantic Current
Labrador Current
Labrador
Canada
Strait of Belle Isle
Atlantic Ocean
Gulf of St. Lawrence
Gaspé Current
Newfoundland
Cabot Strait
Grand Banks
Bay of Fundy
Nova Scotia Current
USA
Gulf of Maine
Cape Cod
Scotian Shelf
Georges Bank
Long Island
Gulf Stream
Bathymetry (meters below sea level)
200
2,000
0
400 km

A polar bear shelters her twin cubs in a dark cleft of the Torngat Mountains, rising from the Labrador Sea.

offshore near Halifax, flowing out to the edge of the Scotian Shelf, offshore of Nova Scotia. But a cold offshoot of the Labrador Current ultimately enters the Gulf of Maine through the Northeast Channel, cooling the waters as far south as Cape Cod.

The Gulf Stream, a warm, salty 100-kilometer- (60-mile-) wide river in the sea, acts as a foil to the Labrador Current's chilling effect. Born in the Gulf of Mexico of the Guiana Current and the North Equatorial Current, the Gulf Stream flows like a warm water jet out through the Florida Straits and follows the coastline until it begins to veer away north of Cape Hatteras, swirling and becoming more turbulent as it heads into the open ocean. Its influence continues to be felt in the Mid-Atlantic region and farther north, however, as pinched-off eddies called warm-core rings drift slowly westward, bringing Gulf Stream waters close to the edge of the continental shelf in the Gulf of Maine and along the Scotian Shelf.

Ecologically, the divide along this grand sweep of coast, from roughly the 60th to the 35th parallel, is Cape Cod. The cape projects into the North Atlantic like a police officer's arm directing marine traffic. South of it, waters are too warm to permanently accommodate animals native to the boreal Acadian zone in the north. North of the cape, cooler waters act as a barrier to animals endemic to the Mid-Atlantic region, or Virginian zone, in the south. This separation of northern and southern species is a consequence of water temperature, which on average is several degrees colder—thanks to the Labrador Current—on the northern shore of the cape than on the southern shore. The landforms north and south of the cape also differ, largely as a result of the difference in the glacial history of the region. In the north, where the Wisconsinan glacier scraped its way across the landscape like a giant bulldozer blade, the contemporary coastline still bears its scars; it is deeply indented, consisting of rocky beaches and cliffs where softer sediments have been largely removed. South of the cape, in the Mid-Atlantic region, where the glacier never reached, a low-lying, mostly linear coastal plain prevails, consisting of sediments eroded from the Appalachian Mountains then molded by the ocean's tides, currents, and storms into sandy beaches, barrier islands, and coastal lagoons.

Far from shore, at the Tail of Newfoundland's Grand Banks, the cold, iceberg-studded Labrador Current, swinging down from the Canadian Arctic and Greenland, and the warm Gulf Stream, curving up from the tropics, vie for supremacy. The first chronicler of New France, Marc Lescarbot, observed this strange conjunction on his first voyage to the New World, in 1606:

> I discovered something remarkable that a philosopher of nature should wonder about. On 18 June 1606 at 45 degrees latitude and at a distance of one hundred and twenty leagues to the east of the banks of Newfoundland, we found ourselves surrounded by very warm water, although the air was cold. Yet on 21 June we were suddenly caught in such a fog that one would have thought oneself to be in January, and the sea was extremely cold.

Lescarbot was witness to the clash of the titans, whereby both of these powerful currents are deflected: the Gulf Stream to the northeast, where it warms northern Europe as the North Atlantic Current; the Labrador Current to the southwest, where it cools the North American coast. While robbing the Northwest Atlantic of a more equable climate, the Labrador Current has bestowed benefits to both wildlife and society. These cold, plankton-rich waters are the foundation of the legendary productivity of Newfoundland's Grand Bank and its more southerly counterpart, Georges Bank, in the Gulf of Maine—two of the richest fishing regions in the world.

*Diverse Waters*

The Northwest Atlantic—the subject of this book—is not a homogenous body of water, which is hardly surprising given its size and reach. As we have already seen, the warm Gulf Stream exerts a moderating influence in the southernmost region, whereas in the north the cold Labrador Current is dominant. Moving from north to south, we encounter water masses that are dramatically different, from very cold, subarctic waters along the Labrador coast to the Strait of Belle Isle, to the cold-temperate waters of the Canadian Maritime Provinces and northern New England, and, finally, to the warmer, temperate waters of the Mid-Atlantic region, between Cape Cod and Cape Hatteras. As a whole, the Northwest Atlantic comprises an ecozone, which is the largest biogeographic unit. South of Cape Hatteras, we enter the Wider Caribbean ecozone—an ocean of warm waters and warm-water species in sharp contrast to the cool waters and cold-water species of the Northwest Atlantic ecozone.

The diverse marine and terrestrial environments within the Northwest Atlantic can best be understood by adopting an ecoregional approach. A useful, if sometimes flexible, concept, an ecoregion is a large, geographically distinct area of land or water sharing a large majority of species and environmental conditions that interact in ways leading to its persistence over long periods of time. It is smaller than an ecozone and larger than an ecosystem. The ecosystem is the basic unit of nature and itself varies in scale. A lake, a forest, and a bog are examples of ecosystems, though a larger geographic unit encompassing all of them might also be considered an ecosystem. Within an ecosystem, living organisms and their environment are inseparable, with a constant exchange of energy and matter, in the form of food, nutrients, water, and waste, occurring between its living and nonliving parts.

Differences in marine environments are more difficult to determine than differences in terrestrial environments. Oceanographic factors, such as currents, tides, water temperature, and salinity, set them apart. Due to the far-reaching, ever-present influence of the ocean, these define the nature not only of the marine life found there but of life along the coast and on land. The types and numbers of organisms are also useful criteria in drawing boundaries.

Using these criteria, the Northwest Atlantic can be divided into five ecoregions. The Mid-Atlantic Bight, between Cape Hatteras and the south shore of Cape Cod, is a marine region guarded by a bastion of barrier islands and penetrated by large estuaries such as the Chesapeake and Delaware Bays. The Gulf of Maine and Bay of Fundy together form a single oceanographic unit that stretches north from Cape Cod along the low, rocky, irregular coast of Maine to the expansive salt marshes and mudflats at the head of the Bay of Fundy. The Scotian Shelf lies off the Atlantic coast of Nova Scotia, where the long reaches of the ocean curls ashore onto beaches of white sand. Sometimes considered together with the Grand Banks of Newfoundland, this area of the continental shelf is described separately on pages 20 and 21. At the heart of North Atlantic coast lies the Gulf of St. Lawrence, a sea within a sea with a shoreline of great topographical contrasts. The North Shore of the gulf presents an "ironbound" aspect of Canadian Shield rocks—a place reviled by its first European explorer, Jacques Cartier, as "the land God gave to Cain"—whereas the southern portion is carved from softer sandstone, creating a welcoming shoreline of wide beaches, barrier islands, coastal dunes, and spits. And finally, in the far north, we come to the dramatic headlands and cliffs of Newfoundland—"the Rock" to its proud inhabitants—and to the forbidding and majestic coastline of Labrador and its iceberg-studded sea. Offshore are the famous Grand Banks, once the greatest cod-producing region on the planet.

Given its size and complexity, it is hardly surprising that the Atlantic is a region of great biological diversity. Offshore, productivity peaks on the shallow fishing banks, where life-giving light penetrates the water column and nutrients are near the surface. The combination of light and nutrients fuels the growth of phytoplankton, the single-celled plants that are the foundation of the marine food web. In turn, the phytoplankton feed the zooplankton, the animal constituent of marine plankton, which include small, shrimplike crustaceans and fish larvae. At the edge of the continental shelf, where waters plunge into the abyssal depths, oceanic fronts concentrate an abundance of zooplankton, which in turn attracts fishes, seabirds, and cetaceans. Inshore, waterfowl and shorebirds exploit the mosaic of mudflats, salt marshes, and rocky shores for their riches. The many islands and towering cliffs that guard the deeply indented northern coastline provide safe nesting grounds for seabird nations, and the barrier islands and lagoons in the south harbor breeding and overwintering populations of shorebirds, waterfowl, and waders. Anadromous species such as Atlantic salmon, alewives, smelt, and American shad run up its rivers, great and small, to spawn, connecting the sea to the hinterland behind the coast.

Located roughly midway between the equator and the North Pole, the coastline of Atlantic Canada in particular serves as a way station for many migratory birds: neotropical land birds making their way to their boreal forest

## COASTAL CLARIONS

HERRING GULLS are ubiquitous along the Atlantic coast. The sight of them scavenging along shorelines and around wharves, or trailing fishing boats in frenzied flocks, and the sound of their piercing, raucous cries, are defining experiences of being on this coast.

Gulls were not always such a dominant presence, however. In 1900, a census of the Atlantic coast recorded fewer than four thousand herring gull pairs—and those were only to be found in New Brunswick and eastern Maine. Gulls were victims of a number of factors. Many fisher-farmer families still occupied the same offshore islands preferred by gulls as breeding grounds, and egging and harvesting of gull chicks were part of a harsh subsistence economy. The pearly gray feathers of gulls were also sought after by the thriving millinery trade in the late 19th century.

Two events—one deliberate, the other fortuitous—converged to reverse the herring gulls' fate. Around the same time most settlers abandoned the bird islands for the convenience of life on the mainland, the Migratory Birds Convention Act was enacted to protect the then-endangered gulls as well as many other bird species. Perhaps the most important factor in the subsequent growth of gull populations was the waste produced by a mechanized fishing industry and an increasingly wasteful consumer society.

Herring gulls are adaptable opportunistic feeders, as are we, and quickly capitalized on these human-generated food sources. As a result, by 1965 the census showed gull populations had ballooned to 100,000 pairs in 240 colonies from New York City to Grand Manan. The breeding population has since spread as far north as Labrador and as far south as northeast South Carolina. That trend has reversed in the recent decade, however, because of the sudden collapse of the groundfish industry as well as the consolidation of municipal dump sites.

The larger great black-backed gulls are found mainly on the Atlantic Coast, while herring gulls, which are highly adaptable, have a breeding range that extends far inland, including every province and territory of Canada and large inland lakes and rivers throughout New England. A large part of their success, while abetted by human factors, is related to their innate behavior. Studies carried out by Nobel laureate Niko Tinbergen demonstrated that gulls are doting parents, boosting survival of their young. Most important, gulls vigorously defend their breeding territory, and during incubation they gently turn their eggs with their bills, a behavior that encourages even development of the embryo. Once the eggs hatch, the gulls immediately remove the broken eggshell, whose shiny

Black-legged kittiwakes are the smallest members of the gull clan. Their commingled cries echo along the Atlantic coast.

white inner surface might attract predators. Parent gulls also guard chicks against attacks by neighbors, since some herring gulls are cannibalistic. Once the chicks begin to explore the feeding territory beyond the breeding ground, they must learn to mitigate the adult's territorial aggressiveness. They do so by assuming a hunched posture, pumping their heads, and voicing shrill calls—the same set of behaviors that they invoke to stimulate the parents to feed them while on the breeding grounds. The behaviors, innate and learned, of both adults and chicks combine to reduce the mortality of the young and contribute to the herring gull's obvious success.

breeding grounds in spring and on their return in late summer, shorebirds migrating between their Arctic nesting grounds and the Southern Hemisphere, southern seabirds escaping the austral winter, and waterfowl moving north and south with the changing seasons in search of open water and food.

The waters and shoreline habitats of the northwest Atlantic are critical to the survival of a number of species. Delaware Bay is a vital staging area for the *rufa* species of red knots, which fly from Patagonia in spring to banquet on the eggs laid by horseshoe crabs that come ashore around the full moon in May, in a wondrous example of the alacrity and fecundity of nature. The outer Bay of Fundy is a critical nursery area for the world's most endangered great whale, the North Atlantic right whale—perhaps only saved from the rapacious whalers of the 19th century by the bay's frequent fogs and treacherous tide-rips. At the other end of the bay, vast mudflats are the feeding grounds for three-quarters of the world population of semipalmated sandpipers before they embark on a nonstop three-day journey over open water to their South American wintering grounds. The estuary of the great St. Lawrence River, coursing out of the interior of the continent, supports the white-winged migration of greater snow geese and the most southerly population of the endangered white whale, the beluga.

But few places, if any, can match the bounteous waters of Newfoundland and Labrador, once home to the greatest feedstock of fish—the Atlantic cod—in the history of human civilization. And despite the modern-day tragedy of the collapse of the northern cod stocks from overfishing, these waters continue to feed a United Nations of seabirds—some 5 million pairs that breed on the islands and cliffs of Newfoundland and Labrador, and 35 to 45 million more that congregate offshore during the nonbreeding season.

### *Coastal and Offshore Habitats*

Within each of these far-flung regions, the major marine habitats occur with varying frequencies and with variations in the native flora and fauna, depending upon the local environmental conditions. Estuaries, large and small, are found all along the coastline wherever a river drains into the sea. By definition, an estuary is a place where fresh and salt waters meet and mix. The term derives from the Latin *aestuare*—"to heave, boil, surge, be in commotion." Most estuaries are highly productive. The turbulent clash of fresh and salt waters underpins this productivity by keeping nutrients in suspension and available to the host of plants and animals that make up the estuarine food web. Two of the major estuaries along the entire Atlantic coast are Chesapeake

The autumnal gold of a salt marsh will soon bequeath its biological riches to the adjacent marine zone.

and Delaware Bays, which receive the outflow of rivers draining from the Piedmont Plateau, east of the Appalachians, toward the Atlantic Coastal Plain. The whole of the Gulf of St. Lawrence, which collects a massive input of freshwater from the Great Lakes, has been described as a large estuary.

The most productive of the marine habitats is the salt marsh. In fact, salt marshes are among the richest habitats on land or sea; their primary production, acre for acre, often exceeds the output of agricultural crops such as wheat and corn. Salt marshes are formed where the tide floods over the land, usually twice a day in accordance with the diurnal tides common to the Atlantic coast, but at least once a month, when the highest tides, the spring tides, occur around the new and full moons. The tides bring with them nutrients in the form of inorganic soils scoured from the sea bottom.

The major sources of the salt marsh's great primary productivity are two salt-tolerant cordgrasses, *Spartina alterniflora* and *Spartina patens*. The former grows on the low marsh, which is flooded daily, and the latter, also known as salt marsh hay because it was once widely harvested for cattle fodder, grows on the high marsh, which is touched by the tide less frequently. Much of this productivity is exported by the tide, or in winter by the ice, as dead plant matter (or detritus) to the adjacent marine zone, where it fuels the offshore food web.

Salt marshes play another critical role in the ecology of the marine zone, acting as nurseries for fishes and invertebrates, many of which, including herring, smelt, and flounder, are of commercial importance. Juvenile fishes frequent marshes, in part because they rely on detritus or the microbes associated with decaying plant matter as food and because of a relative absence of predators. Large salt marshes occur along the shores of Chesapeake and Delaware bays, as well as in the lee of the barrier islands in the Mid-Atlantic region. Salt marshes are also common in the southern Gulf of St. Lawrence, as part of the barrier island and lagoonal system that has developed there, and in the inner Bay of Fundy, where the world-famous tides flood large areas.

The ebbing tides also expose vast tidal flats, or mudflats as they are more commonly called, which in Fundy can be as wide as 5 kilometers (3 miles). Mudflats occur in any areas along the shoreline with sufficient sediment and tidal range and are therefore more common in the sediment-rich areas south of Cape Cod. Organisms like clams and worms, which live buried in the sediment, as well as predators that feed upon them, such as shorebirds at high tide and fishes at low tide, ultimately depend on the productivity of the adjacent salt marsh, or in some cases sea grasses, such as eelgrass, that grow in the sediments themselves. A second and significant source of primary productivity is the microscopic diatoms that coat the mudflats with a living membrane

The hardy community of plants and animals that cling to the rocky coast have adapted to the daily rise and fall of tides and relentless battering of waves.

and whose silica shells act as tiny solar greenhouses. The sediments can vary in size from fine muds to coarse sands, depending on the source of the sediments and the force of the water movements that mobilized and transported them. In the barrier island system, for example, on the seaward side exposed to wind-driven waves, the beaches are sandy, whereas on their less energetic landward side, mudflats and salt marshes develop.

As Mark D. Bertness points out in *Atlantic Shorelines: Natural History and Ecology*, "the critical feature of tidal flats that differentiates them from rocky shores is that many of the conspicuous organisms living in soft substrates are much larger than the particles they live in." The organisms furthermore act as "ecosystem engineers," the worm tubes, grass stems, and mussel beds helping to stabilize the sediments. In other words, they shore up their own houses by the act of living in them.

Rocky shores form a large part of the Atlantic coastline north of Cape Cod, where the glaciers stripped away much of the sediment. What most distinguishes the rocky shores, besides the size of the substrate, is the clearly visible vertical zonation of organisms, an expression of the strong physical stresses on these exposed, wave-battered shorelines. Typical animals and plants have adapted to the environmental stresses at each level of the intertidal zone: lichens in the spray or splash zone above extreme high water; blue green algae and rough periwinkles in the black zone, which is covered only by the highest tides. Below this high-water mark, in descending order, are the barnacle zone, a blazing white due to the limestone shells of the dominant barnacles; the brown algae zone, where knotted wrack and bladder wrack clothe the rocks with luxuriant growth; and finally, in the low intertidal zone, the Irish moss zone, the realm of the red algae. These zones and the organisms are discussed in greater depth in Chapter 5, on the Gulf of Maine, where they have been intensively studied.

Offshore, many islands serve as nesting grounds for seabirds, such as gulls, auks, terns, and petrels, as well as breeding and birthing grounds for seals. Far offshore, we encounter the fishing banks, notably Georges Bank in the Gulf of Maine and the Grand Banks off Newfoundland. The banks stand out from the rest of the continental shelf as shallow areas where sunlight penetrates into the water column and nutrients are found near the surface, a combination that kick-starts the classic marine food web: salts, diatoms, copepods, shrimp, herring—and ultimately, whales, seabirds, and humans. Finally, deep canyons project into the margins of the continental shelf and are home to rarely seen denizens of the deep.

## SCOTIAN SHELF: LIFE AT THE EDGE

THE ATLANTIC coast of Nova Scotia is guarded by nearly four thousand islands, many of which are home to an impressive diversity of breeding seabirds. East of Cape Sable, a series of long inlets furnishes the most important waterfowl wintering areas in eastern Canada, sheltering thousands of Canada geese and black ducks, as well as sixteen other duck species. On Cape Breton's dramatic east coast are two of the most important seabird islands, Hertford and Ciboux, better known simply as the Bird Islands. They are home to some four hundred great cormorants, or 70 percent of the known North American population, qualifying them as the largest breeding colony in eastern North America.

All of the islands rimming Nova Scotia's Atlantic coast have been separated from the mainland only relatively recently by rising sea level and are located close to the shore—with the exception of Sable Island, located 300 kilometers (190 miles) from the mainland. The island is the emergent part of a 1,000-meter-deep (3,280-foot) deposit of sand left behind as an outwash fan when the Wisconsinan glacier melted and sea level rose. Its shallow submerged bars harvested more than five hundred ships in the Age of Sail, earning Sable the epithet "Graveyard of the Atlantic." The crescent-shaped island is also a place of maritime fecundity, notably as the largest gray seal colony in the world, with an annual pup production estimated at between forty and fifty thousand.

Two endemic species are of particular interest: the Ipswich sparrow and the Sable Island pony. First discovered in the winter of 1868 on a dune at Ipswich, Massachusetts, most Ipswich sparrows (a subspecies of Savannah sparrow) breed only on Sable Island. The other endemic species is the famous Sable Island "pony," though "horse" is more descriptive and accurate. Horses were first introduced to the island in 1753 by Andrew le Mercier, pastor of the French Church in Boston, who attempted to establish a colony there. The number of horses in the herd fluctuates significantly, between 150 to 400 animals. During the months leading up to winter, the horses forage on the luxuriant marram grass, achieving maximum fat levels in December. The relatively mild climate of Sable means that the animals do not normally succumb to hypothermia; however, when unusually heavy snowfall covers their food supply, disaster often follows.

Offshore, the edge of the Scotian Shelf (part of the North American Continental Shelf) is an area of enhanced phytoplankton production, which seems to be associated with shelfbreak fronts. An oceanic front is defined as a sharp boundary where currents or water masses with different properties, such as temperature or salinity, collide. At the boundary, there is often increased turbulence,

Atlantic white-sided dolphins plunge through offshore waters, which are host to more than a dozen species of cetaceans.

with downwelling of denser water masses and upwelling of less dense water. Such shelfbreak front areas are ecologically important, as floating organisms, such as jellyfish and zooplankton, collect there and attract swordfish, bluefin tuna, sea turtles, sharks, and seabirds.

Although these fronts may shift positions seasonally, or sometimes daily, they are associated with the continental shelf edge, which is also marked by a series of submarine canyons. The Gully is the largest, plunging to 2,500 meters (8,200 feet) at its deepest point, where it separates the Sable Island Bank from Banquereau Bank. It is unique among shelf-edge canyons of the eastern Canadian margin because of its great depth, steep slopes, and extension far back onto the continental shelf.

What truly sets the Gully apart, however, is the higher density and the greater diversity of whales. The Gully is one of the most important habitats for cetaceans on the east coast of Canada, and perhaps globally. Eight to thirteen species are regularly spotted there. The area is of prime importance for those species that have deep water distributions, especially to the northern bottlenose whale, which has a range limited to the North Atlantic.

### *Coastal Forests and Tundra*

As coast dwellers we often face out to sea, contemplating its mysteries, but we are generally more familiar and more at home with the terrestrial environment at our backs. On land, as in the sea, this Atlantic realm is where north meets south. Three of the global terrestrial communities, or biomes, meet here along a latitudinal gradient—from south to north, the temperate broadleaf and mixed forest, the boreal coniferous forest, and the tundra. Eugene P. Odum coined the term *biome* in his classic text, *Fundamentals of Ecology*, saying that it was "the largest land unit which it is convenient to recognize." A biome is characterized by distinct climax vegetation, which is the final or stable community in a successional series and is self-perpetuating. Climate and the availability of water are major factors determining the characteristics of the biological communities, or ecosystems, typically found there. The community is not defined by vegetation alone, but also by the animals that live there.

*previous spread*: The power of the Atlantic explodes in a geyser of spray.

The great boreal forests that sweep across the northern regions of Canada hold sway in Newfoundland and Labrador, and at the higher altitudes of the Christmas Mountains in New Brunswick, the Gaspé Peninsula in Quebec, and the Cape Breton Highlands in Nova Scotia. But this seemingly limitless coniferous forest gives way to the temperate broadleaf and mixed forest in the Maritime Provinces and northern New England. In northern Labrador, trees of any kind succumb to the subarctic climate and are replaced by sedges, heath shrubs, and grasses.

Regional and local variations in vegetation are found within each of these three major terrestrial biomes. The temperate broadleaf and mixed forest biome, for example, includes a number of distinct ecoregions with characteristic species. In the south, the Middle Atlantic Coastal Forest occupies the flat Atlantic Coastal Plain, from the eastern shore of Maryland and Delaware to just south of the Georgia–South Carolina border. The river swamp or bottomland forests in this ecoregion are dominated by the majestic bald cypress and swamp tupelo. Eastern white cedar swamps also occur as ecological islands along blackwater rivers that snake across the flat terrain. The wetlands of this ecoregion are one of the most important on the continent for diversity of reptile, bird, and tree species, and the remaining pockets of forested habitat are crucial on a seasonal basis for both breeding songbirds and migratory waterfowl. Farther north, the Northeastern Coastal Forest type, typified by white and northern red oak—and formerly by American chestnut—extends over the northern half of Long Island and through Connecticut and Rhode Island to its northern limits in southern Maine. This largely deciduous forest shelters up to 250 breeding birds—the largest number in the temperate broadleaf and mixed forest biome. They include typical southern forest birds such as Carolina wren, red-bellied woodpecker, blue-gray gnatcatcher, worm-eating warbler, and orchard oriole, which are expanding their range northward into Massachusetts and elsewhere in New England.

The Atlantic Coastal Pine Barrens—a very restricted but special forest type—is limited to the sandy, nutrient-poor soils of the coastal plain of New Jersey, the southern half of Long Island, and Cape Cod. This ecoregion supports forests of pitch pine and scrub oaks, which historically were maintained by frequent fires.

The largest of the broadleaf and mixed forest ecoregions is the New England/Acadian Forest, which forms a mosaic of mixed forest types that cover half of New Brunswick; most of Nova Scotia, except for the Cape Breton Highlands; all of Maine, except for the southwestern corner; northwestern Massachusetts; the Champlain Valley of Vermont; and the coastal plain of New Hampshire. Red spruce and red pine are the dominant conifer species, with eastern hemlock and eastern white pine also present, and a mixture of sugar maple, American beech, and yellow birch characterizes the hardwood component. Forest types vary with elevation. Low mountain slopes support

A mosaic of colors reveals the diversity of the New England/Acadian Forest of northern New England and Maritime Canada.

a mixed forest of red spruce, balsam fir, maple, beech, birch, white spruce, and red pine; the valleys contain hardwood forest with an admixture of hemlock in moist ravines; and on the highest mountain peaks, Arctic species can occur as disjunct populations.

Not surprisingly, the Atlantic Ocean exerts a strong influence on the vegetation dynamics, especially in coastal regions, and nowhere more so than along the Bay of Fundy, where strong winds, cooler summers, and shallow soils lead to conifer-dominated forests in which red spruce is prevalent. Along the Atlantic coast of Nova Scotia and Newfoundland, white spruce is the dominant conifer, as it is best able to tolerate salt spray. More sheltered areas bordering the warmer waters of the Gulf of St. Lawrence allow for better growth of hardwoods.

The mixture of hardwoods and conifers attracts an equally rich mixture of bird types, of which 230 breed in the region. These include typical boreal species that prefer coniferous forests, such as sparrows, flycatchers, and woodpeckers, and along the coastlines of the Bay of Fundy and the Atlantic coast of Nova Scotia, typical northern breeding species such as the blackpoll warbler. At the same time, the deciduous forests are a haven for hardwood-loving species such as eastern wood-pewee, black-throated blue warbler, black-and-white warbler, and scarlet tanager. In total, some 468 species have been recorded in the Maritime Provinces and Gaspé Peninsula (the Atlantic Maritime ecozone) alone. More than a third of these are vagrants, however—either land birds blown into the region during their migration by the predominant westerly winds off the continent, or wandering seabirds "wrecked" on land by northeasterly storms over the Atlantic. This same region is home to 58 nonmarine mammal species, and a handful more farther south in New England. Characteristic large mammals include moose, black bear, red fox, snowshoe hare, porcupine, fisher, beaver, bobcat, marten, muskrat, and raccoon.

Neotropical migrants, such as the black-throated green warbler and northern oriole, depend upon the varied habitats of the temperate broadleaf and mixed forest biome.

Typical northern and southern species of birds, mammals, amphibians, and reptiles mix here and in some cases reach the limits of their tolerance. Many northern species such as mink frog, merlin, Wilson's warbler, and moose reach their southern limits in New England, whereas others—including

A string bog mirrors the stark beauty of the boreal forest.

timber rattlesnake, slimy salamander, opossum, and least shrew—fetch up against their northern limits.

The boreal forest, whose climax species are balsam fir and black spruce, occupies the mountainous regions of the Maritime Provinces and northern Maine but has total dominion in Newfoundland and Labrador and on the North Shore of the Gulf of St. Lawrence. Black spruce, along with paper birch and aspen, dominates on disturbed sites, while the salt-tolerant white spruce grows in areas affected by sea spray. The boreal forest is prime habitat for many of the same species found in the New England/Acadian Forest—moose, black bear, and red fox—but it is also home to distinctive species such as lynx and woodland caribou. On exposed headlands, high winds create barrens, covered only by moss-heath vegetation, or cause the krummholz effect, whereby trees suffer dwarfism. Arctic plants appear in disjunct populations, especially on alpine peaks of the Appalachian chain, and Arctic hares appear at the southernmost limits of their continental range in Newfoundland's Long Range Mountains.

Along the northern Labrador coast, steep-sided mountains are punctuated with glacier-carved, U-shaped valleys and fiords extending far inland. By contrast, southern Labrador is gently rolling peatland interrupted by eskers (ridges of sand and gravel that were once the beds of glacial streams), shallow rivers, and string bogs. Stunted stands of black spruce and tamarack are climax species here. White spruce predominates along the coast, but coastal heath covers headlands and cliff summits. The region hosts one of the world's largest herds of ungulates, the George River caribou herd, which crosses the Labrador-Ungava Peninsula between the Labrador Sea and Hudson Bay. This largely unpopulated hinterland is also an important breeding habitat for the endangered eastern harlequin duck, which nests along the fast-flowing rivers, and the peregrine falcon, whose aeries are found in the Torngat Mountains that tower out of the Labrador Sea.

*Waves of Culture*

The fecundity of the North Atlantic was a revelation for the first wave of explorers to reach this continent's shores. A millennium ago, the writers of *The Vinland Sagas* spoke of the "wonder beaches of white sand" and offshore islands of "so many birds . . . that they could hardly walk without stepping on eggs." Five hundred years later, in 1497, the Italian explorer John Cabot spoke of a "sea full of fish which are taken not only with net but also with a basket in which a stone is put so that the basket may plunge into water."

John Smith's exploratory voyages in Chesapeake Bay, from 1607 to 1609, reveal a similar superabundance of fish in the shallow, brackish waters of the Patuxent River estuary, where it meets the bay. Smith's account refers to the "infinite schools of diverse kinds of fish," which included striped bass and bluefish voraciously preying on acres of menhaden, as well as baitfish such as Atlantic silverside, bay anchovy, and mummichog. In addition, spot, croaker, and white perch haunted the tributary creeks, and cownose rays foraged on bottom-dwelling clams and worms.

The earliest natural history of the Atlantic coast was penned by fur trader and fish merchant Nicolas Denys in his 1671 *Description and Natural History of the Coasts of North America*. Geographically, this curious work covers the area then known as Acadia, consisting of the Maritime Provinces of Nova Scotia, New Brunswick, and Prince Edward Island, though his survey begins with Maine's Penobscot Bay. He gives an especially detailed description of the Gulf of St. Lawrence coast—and the mammals, fishes, birds, and trees found there—from Cape Breton Island to the Gaspé, including Prince Edward

Island. It sketches a picture of such a wealth of wildlife that today it is difficult to credit. He named one of his anchorages Cocagne, a place name that persists to this day along the Acadian Shore of New Brunswick. It means "a land of the greatest abundance," which Denys was at pains to describe in his account: ". . . I found there so much with which to make good cheer during the eight days which bad weather obliged me to remain there. All of my people were so surfeited with game and fish that they wished no more, whether Wild Geese, Ducks, Teal, Plover, Snipe large and small, Pigeons, Hares, Partridges, young Partridges, Salmon, Trout, Mackerel, Smelt, Oysters and other kinds of good fish. All that I can tell of it is this, that our dogs lay beside the meat and fish, so much were they satiated with it."

*following spread*: The crazy quilt pattern of fertile farm fields overspreads Prince Edward Island, the "Garden of the Gulf."

A century later, an Englishman gave this wonderstruck account of his arrival on the Grand Banks, after a twenty-seven-day crossing of the Atlantic:

> The 12th of April . . . we came on soundings and knew it to be the Bank of Newfoundland. We would have understood this well enough from other indications for it seemed as if all the Fowles of the Air were gathered thereunto. They so bemused the eye with their perpetual comings and goings that their numbers quite defied description. There can be few places on Earth where is to be seen such a manifestation of the fecundity of his Creation.

Farley Mowat recounts this rapturous testament to the Atlantic's natural bounty in *Sea of Slaughter*, his damning indictment of its squandering in the last three centuries. This destruction of the Atlantic coast's natural capital continues. No part of North America has a longer history of European settlement than the Atlantic coast, and this fact is reflected in the depleted forests and fish stocks and the number of species that are endangered, threatened, or vulnerable—or extinct, as are the great auk, Labrador duck, and sea mink—as a result of habitat alteration or destruction and wanton exploitation.

Human minds have known the Atlantic region intimately, and human hands have shaped it, as demonstrated by the quilted pattern of grain and potato fields on Prince Edward Island, the "Garden of the Gulf," where the original Acadian forest has been all but lost. The first Europeans settled near the coast to be close to the bounty of marine resources—whether fish, whales, oysters, crabs, waterfowl, seabirds, shorebirds, or salt hay.

Today, the northeast coast is the most densely populated region in the United States. More than three-quarters of the residents of the northeast

The densely populated coastal cities and suburbs of the U.S. eastern seaboard meld into a single megalopolis.

coastal states live in coastal counties, an estimated 54 million people from Maine to the tidewaters of Virginia. The region also includes four of the nation's largest metropolitan areas—New York, Washington, D.C.–Baltimore, Philadelphia, and Boston—cities, towns, and suburbs that meld into a single megalopolis.

Adjacent estuaries and coastal waters are currently suffering from eutrophication, or oxygen depletion, as high levels of nitrates and phosphates from agriculture, atmospheric deposition, and sewage are flushed into them. These pollutants may be responsible for the increased frequency and extent of plankton blooms resulting in biotoxin-related shellfish poisoning and the closure of the region's beaches to swimming because of elevated bacterial levels, and may even be responsible for mortalities in marine mammals.

A quarter of sediments near major urban areas contain unacceptably high levels of contaminants, and 10 percent of fish have elevated levels of contaminants in their edible tissues. Contamination with heavy metals, such as zinc, cadmium, copper, lead, and nickel, is highest in southern New England, reflecting the high density of population and intense industrialization of the region. In the same region, suburban sprawl has eaten all but 2 percent of the original terrestrial habitat. Even so, the marshes, lagoons, barrier islands, and offshore waters within sight of this crush of people, asphalt, and concrete continue to be host to large numbers of seabirds, waterfowl, shorebirds, and waders.

At the same time, far from the eastern seaboard's bright lights, vast areas, such as the hinterland of Labrador's interior and its 20,000 kilometers (12,400 miles) of coastline, are known only to Aboriginal peoples—a succession of Inuit and Indian cultures—who sparsely inhabited these lands long before European contact. One of the oldest (10,600 BP) and richest Paleo-Indian sites in eastern North America is located at Debert, Nova Scotia, at the head of the Bay of Fundy's Minas Basin. Although the continental glacier had long disappeared, archaeologists believe that local glacial remnants remained lodged in the nearby hills and that the vegetation was typical of tundra. These early peoples probably camped here to exploit the seasonal migration of woodland caribou, which were extirpated from Nova Scotia only in the 1920s.

Later peoples turned to the sea to make their living. This so-called Maritime Archaic culture is best known from Newfoundland and Labrador, where it may be traced back more than eight thousand years, but Memorial University of Newfoundland archaeologist James A. Tuck believes that it embraced the entire Atlantic coast from northern New England to northern Labrador. They had developed a very sophisticated sea-hunting technology, which included barbed and toggle-type harpoons. Sites in northern Maine, dating to 4,500 BP, include bones from swordfish, gray and harbor seals, walrus, codfish, sturgeon, various seabirds, and, at one site, the extinct sea mink. They also harvested land mammals—primarily deer, moose, bear, and beaver—using lances with bone and ground-slate tips.

At about the same time, farther south in Chesapeake Bay, Native Americans began exploiting a variety of shellfish, including oysters, softshell clams, marsh periwinkles, and ribbed mussels. They also began planned burning of the forests, a practice that boosted browse plants for white-tailed deer, which furnished most of the meat in their diet.

There are wild places that, although accessible, remain essentially unknowable, such as the great Bird Rock at Cape St. Mary's, Newfoundland, divided from human contact by an aerial chasm both physical and experiential. In *Wild America*, no less a naturalist than Roger Tory Peterson and his companion British birder James Fisher enthused about the spectacle that they beheld there:

## PEOPLE OF THE BAY: *Powhatan to Today*

WHEN THE English adventurers led by John Smith arrived in Virginia in 1607, they found an already well-established culture: the Powhatan Confederacy. This Algonquian-speaking people occupied the land surrounding the tidewaters of Chesapeake Bay, living in small villages of forty to two hundred persons and in seasonal hunting camps. Most villages were located along the floodplains of the region's major rivers, the James, York, and Rappahannock, where the rich alluvial soils fertilized their crops of corn, beans, and squash, and the location on low terraces gave them easy access to the waterways with its rich resources. Without this well-established food-producing system of bread, corn, fish, and game, the newcomers, by their own admission, "had all perished."

The indigenous people of the Chesapeake region practiced shifting cultivation, clearing new fields each year and moving their sapling-and-mat houses when fertility of the old fields ran out. The houses were durable structures made of marsh reeds woven and sewn into mats stretched over bent saplings at least 5.5 meters (18 feet) in length. These oblong dwellings housed extended families in the agricultural villages, and circular houses for individual families were also constructed as winter hunting quarters and moved as needed.

The natives of coastal Maryland and Virginia had access to a large and varied larder of natural foods. Some 1,100 edible wild species have been identified. Two of the most important were tuckahoe (arrow arum) roots as a source of starch and greenbrier leaves for greens. Freshwater marshes on wide floodplains of the region's meandering rivers were virtual breadbaskets, producing tuckahoe as well as the starchy seeds and tubers of pickerel weed, wild rice, and spatterdock, or cow lily. Abandoned fields sprouted blackberries and raspberries, which lured game animals such as raccoons and opossums.

In many areas the deciduous forests grew to the waterside. In autumn the many nut-producing trees yielded acorns, walnuts, beechnuts, hazelnuts, and chestnuts, and these in turn attracted a variety of game, including deer and wild turkeys, staples in the native diet. Fowling for geese and dabbling ducks was an important winter activity. Fish weirs constructed of stones and pole-and-reed fences might be operated year round, but were most productive in spring when anadromous fishes, such as shad and alewives, were running the rivers. If all else failed, oystering in the bay's shallow waters or clamming along the beaches of the Atlantic shore and the back barrier bays staved off hunger. Middens indicate that Chesapeake's famous blue crabs were also a favored food.

When John Smith arrived on the scene, perhaps fifty thousand Native Americans—Algonguian-, Iroquoian-, and Siouan-speaking—lived in the Chesapeake

The bounty of Atlantic cod first lured Europeans across the Atlantic and for five centuries sustained a commercial fishery and fishing culture.

Bay region. Within a century, European settlers in Virginia and Maryland outnumbered natives. Africans arrived, first as indentured servants and later as slaves. Today the bay is surrounded by 16 million inhabitants, a number expected to reach nearly 18 million by 2020. The most important factor affecting the Chesapeake environment has been the post–World War II population boom, which witnessed a doubling of bay residents. Although parts of the precolonial ecosystem remain intact, it has been drastically altered by urbanization, agriculture, and the denuding of the original forest. The excess nutrients and sediments pouring into the bay have had an overwhelming impact on this once-clear-water aquatic habitat, clouding its future.

Several Indian tribes recognized by the Commonwealth of Virginia still live along the river valleys and shores of the bay. As well there are Indian people in southern Maryland, living without a reservation, and others on the Maryland Eastern Shore. Descendants of the Powhatan started shad hatcheries in the early 20th century, and today modern hatchery laboratories on both the Pamunkey and Mattaponi reservations produce millions of young fish annually. This has led to a resurgence of an important food and sport fish decimated by four hundred years of pollution and damming of the bay's rivers.

Circumventing the noisy kittiwake slab we finally reached the gannets, whose great stack (350 feet) is connected with the 400-foot mainland cliff only by a low impassable ridge. What a wonderful show they put on for us! Thousands of birds, with a wingspread of six feet, covered the top and sides of the stack.

I have experienced the same exhilarating feeling at the Cape St. Mary's gannetry, which has grown in size since Peterson visited it in 1953. Both on land and at sea, however, the vitality and composition of ecosystems have been altered by four centuries of resource exploitation, primarily forestry and fishing, but also farming, and by burgeoning population growth and the accompanying loss and pollution of inshore habitat. The pattern first documented in Europe has been repeated on this side of the Atlantic: depletion of freshwater resources, followed by decimation of the inshore and eventually the offshore resources—a trend that inevitably landed Spanish, Portuguese, French, and English fishers and whalers on North American shores. The damage caused to the original ecosystem of the North Atlantic is likely to be exacerbated by climate change and its effects, including sea level rise, which is already significantly altering the shape and character of the Atlantic coastline.

Although much has already been lost, much of the natural heritage of the region has also been retained and, in some cases, has even been recovered in recent decades. For example, one hundred years ago, three-quarters of New England was cleared and under cultivation; today, the opposite is true, with forests covering 75 percent of the landscape. Looking seaward, the picture is less encouraging: recent studies carried out by Dalhousie University marine scientists in Halifax, on the doorstep of the Atlantic, show that, worldwide, the populations of all large predatory fishes—swordfish, salmon, and tuna, for example, as well as whales, sharks, and sea turtles—fell by 90 percent within fifty years of the onset of industrialized fishing. The numbers of northern cod,

which so impressed John Cabot, are even bleaker, as the current biomass is only 10 percent of historical stocks. At the same time, some seabird populations, as well as some seal and whale species, have been brought back from the brink of extinction by concerted conservation efforts. Whether the Atlantic can regain its former fecundity will depend, in large measure, on our continued commitment to changing our own behavior as the ocean's beneficiaries and stewards.

The great seabird colony at Cape St. Mary's, Newfoundland, closes the gulf between human experience and the wonders of the natural world.

# OCEANS AND MOUNTAINS

## *The Geology and Paleontology of the Atlantic Coast*

AT GROS MORNE National Park, on the west coast of Newfoundland, a great chunk of the Earth's innards is exposed on the surface. This mélange of oceanic crust and underlying mantle—a complex known as ophiolite—is completely infertile laden as it is with heavy concentrations of nickel and iron and, in the words of nature writer Wayne Grady, most resembles the "tailings from some colossal iron ore mine." It has proved impervious to erosion, having not broken down into soil in the past half billion years. This slab from deep within the Earth now stands nakedly on the surface, an enduring monument to the process that has shaped and is continuing to shape the planet.

It has been said that Gros Morne is to geology what the Galápagos is to biology. Just as the Galápagos Islands, with their famous finches and tortoises, seeded and supported Darwin's theory of evolution by natural selection, so Gros Morne's rocks are testament to the unifying theory of the evolution of Earth's continents and oceans—the theory of continental drift first proposed by the German meteorologist Alfred Wegener in his seminal book, *The Origin of Continents and Oceans*, published in English in 1924, and later refined as the theory of plate tectonics.

< The Tablelands of Gros Morne National Park, Newfoundland, rise from the sea as dramatic testament to the theory of plate tectonics.

Wegener, a German-born Arctic explorer and meteorologist, noticed that the coastlines of Africa and South America seemed to dovetail, "as if we were to refit the torn pieces of a newspaper by matching their edges." He surmised that they had been a single landmass and, over the eons, had somehow drifted apart. He set about compiling other evidence for his theory, matching coal deposits and fossils found on either side of the oceans where the continents had presumably been joined. One such fossil was of the fish-eating freshwater reptile *Mesosaurus* which lived only in Brazil and southern Africa about 270 million years ago. But the scientific community vehemently rejected Wegener's radical theory of continental drift before his death on an expedition to Greenland in 1930, and it would be nearly half a century before it was vindicated. What was missing was a mechanism to explain how continents could drift and collide and how oceans could open and close.

The new theory of plate tectonics, first put forward by a University of Toronto professor, J. Tuzo Wilson, in 1965, solved that problem. It described a dynamic view of the Earth in which a series of rigid plates—some twenty of them on the planet's surface—are in constant slow motion, migrating at the rate one's fingernails grow. Over millions of years, they do eventually bump into each other: the results can vary but are always dramatic—earthquakes, volcanic eruptions, the creation of islands, or a monumental crumpling of the landmasses at their leading edges, which pushes up mountains. And then they drift apart again, creating separate continents and oceans in between. A record of this process is uniquely preserved at Gros Morne in the form of ophiolite.

To understand plate tectonics, it is first necessary to examine the structure of the Earth, from the inside out. The Earth has been compared to a series of nested balls. At the center is a solid core, 2,740 kilometers (1,700 miles) in diameter, surrounded by a 2,000-kilometer (1,200-mile) thick outer core of molten nickel and iron. The core, in turn, is surrounded by the mantle, a 3,000-kilometer-thick (1,900-mile) layer of dark, coarsely crystalline rock called peridotite, which is rich in both magnesium and iron. Floating atop the mantle and core is the relatively thin crust, which varies in thickness from 6 to 35 kilometers (4 to 22 miles), being thicker under mountain ranges than under oceans.

The crust is formed of lighter minerals that floated to the top of the mantle early in the Earth's formation. There are two types of crust: oceanic and continental. Oceanic crust is similar in composition to the mantle, consisting of iron- and magnesium-rich rocks like basalt. Continental crust is formed

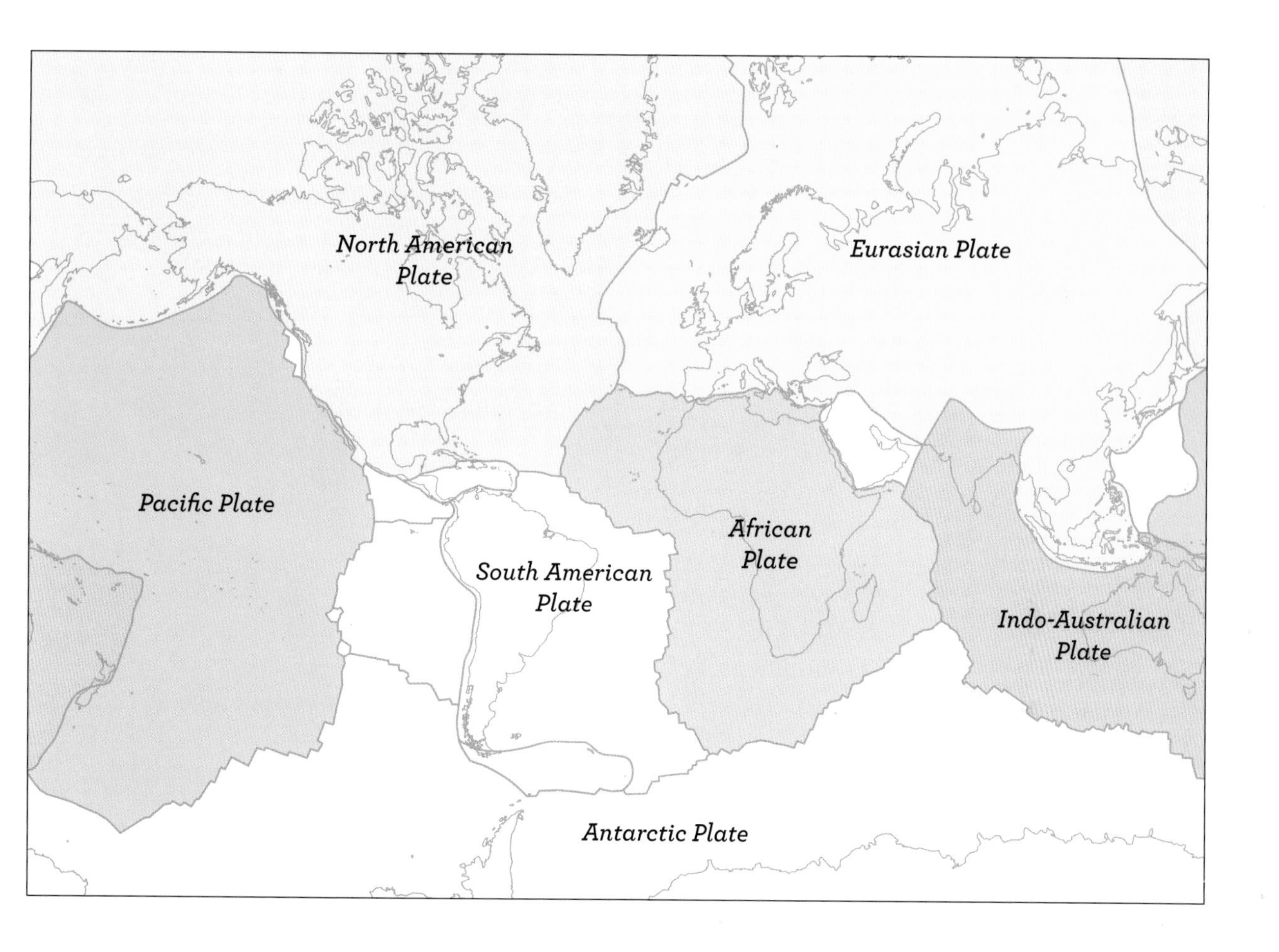


of granite and other igneous and metamorphic rocks with a higher silica content and is lighter than its oceanic counterpart. This difference is important in understanding the dynamic process of plate tectonics.

This traditional view of the Earth's inner structure was based largely on composition. More recently, geophysicists have devised a descriptive system based on the strength and fluidity of the layers. In this system, the outer layers of the Earth have been divided into lithosphere and asthenosphere. The lithosphere is defined as the cooler, more rigid outer layer, which includes the crust, both oceanic and continental, and the upper mantle. The asthenosphere is the weaker, more plastic layer underlying it.

According to plate tectonics, the lithosphere is divided into a jigsaw puzzle of twenty separate pieces or plates that are propelled, like colossal barges, by convection currents within the plastic asthenosphere. Resulting from the temperature differences in the molten rock, these currents have been compared by science writer and geologist Simon Winchester, in *Krakatoa: The*

MAP: The outer layers of the Earth are divided into twenty tectonic plates, which, like colossal barges, are in constant motion relative to one another. Seven of the major plates are identified here.

*Day the World Exploded: August 27, 1883*, to "a vat of vegetable soup simmering on the stovetop." Convection currents rise up from hot regions, as deep as 1,500 kilometers (930 miles), at the slower-than-molasses rate of millimeters per year. They pass through the pliable asthenosphere, where they encounter the brittle crust and then turn back down again. In places, however, this magma breaks through the crust and bubbles to the surface, resulting in volcanic eruptions on land and under the oceans.

Today a series of volcanic islands in the Mid-Atlantic stretches from Iceland in the north to Bovet Island near the Antarctic ice pack. These islands mark the Mid-Atlantic Ridge and are the only emergent part of a 16,000-kilometer-long (1,000-mile) mountain chain. The Mid-Atlantic Ridge occurs at a plate boundary, separating the Eurasian Plate from the North American Plate, where the seafloor is spreading as the molten rock wells up and spreads out and is then replaced by more molten rock. Compelling evidence for seafloor spreading surfaced in the 1950s when magnetic readings, obtained by the U.S. Coast Guard, revealed that the farther from an oceanic ridge you go, on either side, the older the rocks.

*facing page*: Killdevil Mountain, Newfoundland, is part of the Appalachian mountain chain thrust up when continental plates collided 400 million years ago.

The oceanic crust acts much like a conveyor belt. New crust is formed at oceanic ridges by upwelling of magma but is destroyed when it reaches the edge of another plate. This latter area is a subduction zone, where one plate slides beneath the other and is melted and incorporated again into the molten layer underneath. (The addition and subtraction of crust is, in this way, kept in balance, and therefore the Earth itself does not expand in size.) Currently, the spreading of the sea floor at the Mid-Atlantic Ridge is causing the Atlantic Ocean to widen by roughly 2 centimeters (less than an inch) per year. The opposite process is happening in the Pacific, where the ocean floor is being swallowed under the North American Plate.

The plates, which can consist of both oceanic and continental crust, are in constant motion relative to one another in a kind of millennial bump and grind. The area where two plates meet is called a plate boundary, and a number of things can happen there, depending upon the nature of the two plates. If a plate of oceanic crust meets one of continental crust, the heavier oceanic plate is subducted. If two plates of entirely oceanic crust meet head-on, a subduction occurs randomly, with either plate being subducted. This encounter is often accompanied by the creation of an arc of volcanic islands, such as the Aleutian Archipelago. But if the plates are composed mostly of continental material, subduction may not occur at all when they collide. They may, instead, crumple at their edges and set off a mountain-building episode, also known

Mistaken Point, Newfoundland (top), preserves an ancient deep-sea community of soft-bodied, multicellular organisms (bottom), the Ediacarans, variously described as ostrich feathers, Christmas trees, or spindles.

## OSTRICH FEATHERS, CHRISTMAS TREES, AND SPINDLES

PRESERVED IN the fine volcanic ash-beds at Mistaken Point, at the southeastern tip of Newfoundland's Avalon Peninsula, is a curious community of bottom-dwelling creatures, the Ediacara, which inhabited the Iapetus Ocean 575–560 million years ago. Ediacarans are the earliest known complex multicellular organisms, and those at Mistaken Point are the only deep-water marine fossils of this age anywhere in the world. Buried instantly, they are preserved in exquisite detail in dark siltstone and sandstone exposed by the battering of the Atlantic's waves. These soft-bodied animals are variously described as resembling ostrich feathers, Christmas trees, spindles, or merely flattened blobs and discs imprinted with intriguing filigrees of creases and branches. They appear to have had neither head nor tail; nor are there obvious circulatory, nervous, or digestive systems.

The Mistaken Point Ediacarans appeared at that critical time in Earth's history when ecosystems dominated by multicellular animals were supplanting those dominated by microorganisms. Some were sedentary, attaching themselves to the seafloor by a holdfast, much as barnacles or kelp do; others may have floated free. Recently discovered specimens of *Charnia,* at Mistaken Point, had holdfasts and fronds nearly 2 meters (6.5 feet) in length, making them perhaps the oldest large, morphologically complex fossils known anywhere.

It appears that most, if not all, Ediacarans disappeared just before the Cambrian explosion of multicellular organisms. Many shelly creatures, including trilobites, emerged during this era, and these new Cambrian predators may have found the immobile Ediacarans easy pickings. The spectacular preservation of whole Ediacaran communities at Mistaken Point provides a snapshot of an ancient deep-sea ecosystem, which in many respects resembles a modern sea slope community in its diversity, abundance, and spatial patterning. Furthermore, it suggests that the deep sea was colonized quickly at this early stage of animal evolution.

coast of North America is known as a collision coast, and much of the geological complexity found there is due to the folding and faulting that accompanied that collision of continents.

The Appalachians were actually built in two stages. The first event, the Taconic orogeny, is named for the Taconic range of mountains of western New England and southeastern New York State. About 450 million years ago, in the Late, or Upper, Ordovician period, the northern part of Europe drifted northwestward, closing the ocean gap between it and the North American-Greenland plate of Laurentia. The suture or area of contact can be traced from the eastern side of Norway, through southern Scotland, across Ireland to the central part of Newfoundland. Eventually the sediments that had been deposited along the northern part of North America were crushed against the margin and thrown up to form a 2,500-kilometer (1,500-mile) chain of mountains that arced from western Newfoundland through eastern Quebec and southward through Vermont to eastern New York State.

*following spread:* Rocks rise from the sea near Gros Morne, Newfoundland, which has been called the "Galápagos of Geology."

Then, for the next 75 million years, things went quiet on the margin of the North American Plate. The continents were still moving inexorably toward each other, however, resulting in volcanism that spewed out lava flows and ash deposits during the Silurian Age (430–395 million years ago) and left a blanket of deposits 5,000 meters (16,000 feet) deep around Yarmouth, Nova Scotia—including at Cape Forchu, where I played as a child—and northern Maine. During the Devonian period that followed, 390 million years ago, the continental plates of North America and Europe squeezed closer together, further folding the continental sediments and forming the Euramerica continent. This was the Acadian orogeny, the second phase of Appalachian mountain building. As a result of this episode, the mountain chain was stretched from Newfoundland south to Alabama. Parts of the range, like the Gaspé, were unaffected by this second cataclysm. But new parts were created, such as the Nova Scotia uplands, and other parts, like the New Brunswick highland and uplands of Newfoundland, were refolded by this second event.

As the Earth was being altered by these titanic events, so life on the planet was undergoing radical changes. The fossil record of these evolutionary advances is truly remarkable along the Atlantic coast, where the ocean itself has opened wide windows onto time's parade of organisms. At various places along the coast, we can witness the development of the first vertebrates, the fishes, and the evolution of life onto land, with the rise first of the amphibians and later of the reptiles, the first truly terrestrial group of animals.

*Out of the Water*

During this mountain-building period, life was evolving in the planet's waters and was about to take its first tentative steps onto land. Diversity increased during the Silurian (443–418 million years ago) as a variety of invertebrates, including brachiopods, clams, snails, sea lilies, bryozoans, and trilobites populated marine waters. During the Devonian (418–362 million years ago), new species of flora and fauna proliferated in the water and on land. But it is rightly known as the Age of Fishes, because during the 56 million years of the Devonian many new fish prototypes appeared. The best record of this proliferation of fish species is found in the Miguasha cliffs at the mouth of the Restigouche River, which is now famous for its large Atlantic salmon. The pioneering geologist Abraham Gesner, best known for inventing kerosene, first discovered the wealth of ancient fish fossils preserved at Miguasha in 1842, and the cliffs were made a UNESCO World Heritage Site in 1999.

PLACODERM

It now appears that Miguasha was part of a large drainage system on the southwest margin of the Euramerican continent, which was flowing into the Rheic Ocean. The young Appalachians towered above it and were shedding their sediments into it, forming an impressive delta that may have rivaled that of today's great river basins, such as the Amazon, Mississippi, and Nile. During the Devonian, the warm estuarine waters at Miguasha were overflowing with fish species, large and small.

Devonian fish species fall into two large groups: agnathans, or jawless, fishes, and the gnathostomes, the jawed fishes. Only two of the agnathan group have survived: the lampreys and hagfish. These eel-like fishes have sucker-like mouths furnished with a rasp. Lampreys attach themselves to other fishes, like salmon, which they feed on parasitically, whereas hagfish scavenge dead and dying animals, such as whales that fall to the ocean bottom. The rest of the jawless fishes disappeared at the end of the Devonian, as well as the placoderms, a formerly dominant fish type that were armored with bony plates and had primitive jaws.

Three types of jawed fishes survived. Among them were the sharks and rays (chondrichthyans), which thrive to this day. Another was the ray-finned fishes (actinopterygians), which, at 29,000 species, is the most successful vertebrate group on the planet. They were the first true bony fishes, with a bony vertebral column rather than the cartilaginous one that sharks and rays have. But perhaps the most important group, from an evolutionary viewpoint, is the lobe-finned fishes, which gave rise to the first four-legged, air-breathing terrestrial vertebrates, the tetrapods. The highest number—some 3,000—and best-preserved specimens are known from Miguasha. The most important of these lobe-finned fishes, *Eusthenopteron foordi*, has been nicknamed the Prince of Miguasha.

It has also been called "the fish with legs," because its fin bones are so similar in structure to those of the four-legged tetrapods, including amphibians, which would soon adapt to living part of their life on land. (Subsequently, reptiles, birds, and mammals would also join the tetrapod assemblage.) This primitive fish shared other traits with tetrapods, having teeth made of folded sheets of dentine and, perhaps most intriguingly, internal nares (called choanae) opening into the mouth, which would have allowed it to breathe air. It is not known for sure that it was an air breather, but examination of the skeleton indicates that "the Prince" quite likely had lungs. It also sported sail-like fins far back on the body, which allowed it to accelerate rapidly in pursuit of prey. As a top predator, it was equipped with formidable fangs, which it obviously put to good use, as whole fish have been found fossilized within its abdomen, including those of its own species.

*EUSTHENOPTERON*

It is its lobe fins, however, that mark its appearance as a milestone in the story of evolution. *Eusthenopteron* had a powerful pair of pectoral fins that closely match the bone structure in the arms of early tetrapods. The fins featured an upper arm bone (humerus), two forearm bones (radius and ulna), and primitive wrist bones. Modeling experiments show that it could swing its muscular forelimb back and forth in a 20- to 25-degree arc. Most of the movement was at the shoulder joint, but there was some flexibility at the elbow. It was not able to raise itself off the ground and support its own weight as later amphibians could do, but it could drag itself through the soft mud between drying pools in a primitive walking movement and in doing so take a first tentative step onto land.

By the end of the Devonian, invertebrates were moving onto the land. It is likely that leeches, flatworms, and earthworms had come ashore, even though there is no fossil evidence of their presence. Snails and insects also probably were now terrestrial. Fossil evidence discovered near Dalhousie Junction in northern New Brunswick demonstrates that a centipede-like creature *Eoartheopleura* was crawling on land, as was an air-breathing scorpion. These two fossils are the earliest evidence of life on land in North America. It is not surprising that they were arthropods, equipped to cope with water loss and the need for structural support—two of the main challenges of life on land.

Life on land was made possible in the first place by the adaptation of plants, the source of life-giving oxygen. At first, plants kept their roots in the water

Horsetails are diminutive descendants of the giant club moss trees of the Carboniferous, or Coal Age.

and grew above the surface like reeds. Gradually they moved away from the water's edge, and by the end of the Devonian, 10-meter-tall (30-foot) trees had evolved. These were ancestral forms of the horsetails, club mosses, and ferns, which were to assume truly giant forms in the Carboniferous (262–290 million years ago).

THE CONTINENTAL PLATES continued to coalesce, until a third megacontinent, called Gondwana—incorporating Australia, Antarctica, South America, Africa, India, southern Europe, and Florida—drifted into place and melded with Laurentia. This formed a supercontinent that Wegener called Pangaea, or "one land," which became the starting point for his theory of continental drift. Pangaea would stay intact for the next 100 million years.

As the continental plates closed and the mountains rose, a series of faults formed that produced deep basins where parts of the Earth's crust subsided. Erosion of the nearby mountains, meanwhile, poured masses of sediments into these basins, which accumulated to thicknesses of several kilometers. The Maritimes Basin extended from central Nova Scotia northward across the Gulf of St. Lawrence to the shore of the present-day Gaspé Peninsula and from western New Brunswick to Newfoundland. Rivers and streams deposited great loads of gravel, sand, and mud into this basin. In some places, deep lakes were formed and in other places the rivers fanned out to form deltas. The muddy environments of these ancient lakeshores and river deltas were ideal for the preservation of fossil trackways.

In 1841, Sir William Logan, who was to become the first director of the Geological Survey of Canada, was intrigued by some building stone on the wharf at Windsor, Nova Scotia. His inquiry about its origins led him to Horton Bluff, overlooking the Avon River estuary. There he discovered a set of footprints made by an amphibian 350 million years earlier. It was a radical discovery, because, until then, it was believed that fishes were the only vertebrates of the Carboniferous period and that they did not crawl onto land until the Permian period, some 50 million years later. Logan's discovery furnished the first evidence of a tetrapod's successful foray onto terra firma.

In 1964, some 120 years after Logan's groundbreaking discovery, an even more impressive set of trackways was uncovered at Horton Bluff. The twenty-seven footprints spanned a 20-meter (65-foot) stretch of beach, formed of 350-million-year-old sediments that showed distinct ripple marks made by wave action on a shallow lake bed. At the time, they were the oldest vertebrate footprints in the fossil record, and they were of unprecedented size—0.3 meters (1 foot). They were also deep, with raised edges where the trackmaker had sunk into the mud. Obviously they had been made by a large, heavy animal. The absence of claws and the width of the trackway marked the creature as an amphibian. No bones large enough to match the tracks were found, but a likely candidate for the trackmaker was an embolomere, an order of extinct amphibians more related to crocodiles than to living frogs and salamanders. It probably had fangs and was the most feared predator of the Carboniferous swamp. The size of the fossil also meant that amphibians had already been around for a considerable time.

Amphibians did not actually rule the Earth, being only partially adapted to life on land; they still required water to breed, just as frogs do today, laying and fertilizing their eggs externally in water. But amphibians were the dominant vertebrate group during the Carboniferous period.

THE VARIOUS CONTINENTAL plates had now fully merged, and land plants had fully taken hold, especially in the equatorial regions, which included much of the eastern seaboard. The giant club moss trees, such as *Sigillaria* and *Lepidodendron*, grew 30 to 40 meters (100 to 130 feet) tall. Giant tree ferns flourished in the understory, and bamboo-like horsetails, *Calamites*, fringed the riverbanks. Most plants bore spores rather than pollen. There were also many seed-bearing plants (gymnosperms) spreading across the land, as well as the ancestral conifer, *Cordaites*, that broadcast its seed as pollen. When fossilized, this luxuriant plant growth furnished the coal beds that give their name to this geological period—the Coal Age or Carboniferous period.

These lush Coal Age forests often grew in lowland swamps and along riverbanks, which were subject to periodic flooding. Such inundations buried the forests in sediment, where, over time, they were compressed into coal seams. As living forests, these environments were home to a host of invertebrates that reached gigantic sizes. The giant arthropod *Arthropleura* grew to 2 meters (6 feet) in length and left its caterpillar-like tracks in the muds of the forest floor, where it fed on the rich supply of dead wood. It probably most resembled an oversize sowbug, or wood louse, and played an important role in recycling the abundant litter on the forest floor. As the largest land creature in the Coal Age forest, it likely had few enemies. Its aerial counterpart was a giant dragon fly, *Meganeura*, whose nearly 1-meter (3.2-foot) wingspan was comparable to that of modern birds of prey. This ancient insect was equipped with formidable mandibles, or jaws, and it could swoop down and pick off its prey in its long, basket-like legs. All of these Coal Age giants are represented in the coal beds at Joggins, Nova Scotia, which was recently given World Heritage site status for the preservation of a Coal Age ecosystem in its seaside cliffs along the shores of the Bay of Fundy.

*ARTHROPLEURA*

The most significant creatures to come out of the Coal Age, however, were diminutive—the first reptiles. A record of the emergence of this animal group is uniquely preserved at Joggins, which Abraham Gesner described as "the place where the delicate herbage of a former world is now transmuted into stone."

The importance of this area to our understanding of the Coal Age and to the evolution of life on Earth was sealed in the mid-19th century when the Scottish-born founder of modern geology, Sir Charles Lyell, first visited. He had come to see the fossil tree trunks that are preserved in the upright growing position in the cliffs. (The fact that the trees were fossilized in the position in which they had grown contributed to the understanding of how coal itself was formed, confirming that coal beds developed in place from a

The Joggins Fossil Cliffs preserve a complete Coal Age ecosystem, including the oldest reptiles in the geological record.

living forest rather than from dead plant material that had drifted into place—a common misunderstanding of the time promulgated by no less a luminary than Charles Darwin.) Lyell was duly impressed, and stated in a letter that the forest of fossil coal trees was "the most wonderful phenomenon perhaps that I have seen," and that "this subterranean forest exceeds in extent and quantity of [fossil] timber all that have been discovered in Europe put together." Lyell was so taken with the degree of preservation at Joggins that he returned to the site a second time, a decade later, in 1852, in the company of the young Canadian geologist Sir John William Dawson. On this occasion, what they discovered inside one of these fossil tree trunks would provide insight into the evolution of the first truly terrestrial creatures—reptiles.

These early reptiles differed from their amphibian forebears in subtle but significant ways. They had comparatively small heads, one-fifth of the trunk length rather than the one-third to one-quarter typical of amphibians. The head was held higher, and they had stronger jaws. Overall, they had more lightly built skeletons, which aided their mobility on land. Reptiles also have scaley, waterproof skins that prevent evaporation of water, an important adaptation to surviving in the air.

But the most important innovation was their novel reproductive strategy. Amphibians were tied to the water, as frogs and salamanders are today, by their need to lay eggs there; protected only by a thin membrane, their eggs would dry out if not immersed in water. Reptiles and their descendants, the birds and mammals, were amniotes, meaning that a membrane (the amnion) protects the embryo inside a closed egg. Amniotes laid eggs with a leathery or hard calcereous shell that was semi-permeable, allowing for the exchange of air and gases. Inside was a complex of protective membranes and fluid, as well as a food supply in the form of a yolk sac—all that the developing embryo needed to survive until it hatched. The amphibians also retained a fishlike form of fertilization by shedding sperm near a mass of freshly laid eggs, a wasteful technique that nevertheless works well enough in water. Reptiles, however, used internal fertilization, or penetrative sex, a more economical way to deliver sperm.

The early reptiles that emerged in the Late Carboniferous were small, slender animals. The one that Lyell and Dawson found preserved inside a fossilized tree trunk, *Hylonomus lyelli*, was a mere 20 centimeters (8 inches) from head to tip of tail. One hundred million years later, these diminutive creatures gave rise to the great dinosaurs. Again, some of the oldest fossils of these "terrible lizards" are found along the shores of the Bay of Fundy.

## STRANGE REPOSITORY

THE HOLLOW trunk of a lycopod tree at Joggins, Nova Scotia, yielded what is still the oldest reptile in the fossil record, dating to 300 million years ago, in the Pennsylvanian period of the Carboniferous. The lycopods are giant relatives of the living club mosses and grew along river channels on a flood plain. Waters washed down from nearby highlands and crested the natural riverside levees, depositing sediments over the roots of the trees, thereby killing them. A windstorm eventually snapped off the trunks, and the soft inner tissues of these primitive trees rotted.

*HYLONOMUS LYELLI*

The bark was durable, however, creating a hollow stump, which Sir John William Dawson dubbed a "strange repository." With each new flood, sediments accumulated around the base of the dead tree until it reached the lip of the trunk. At this point, the primitive reptile either fell or crawled into the tree trunk, where it was preserved when another spate flooded the area. The creature, which Dawson named *Hylonomus,* or "forest dweller," must have been a terrestrial animal to have been preserved in this manner. The skeletons are mixed with charcoal, suggesting that the reptile may also have entered the trunk through fire scars. The presence of abundant coprolites, or fossilized feces, also suggests that the reptiles may have used the hollowed-out tree trunks as dens where they moved in to feed on millipedes, insects, and snails (also preserved in the trunks) and eventually became trapped. Either way, the fossilized skeletons themselves confirm that this was a new type of animal, a reptile rather than an amphibian.

### *The Dawn of the Dinosaurs and Birth of the Atlantic*

Following the Carboniferous, in Permian times (290 to 248 million years ago), the continents continued to press closer together. The area of the contemporary east coast of North America was in the region of the tropics but was also continental—that is, stuck in the interior of the supercontinent Pangaea. The coming together of Pangaea caused profound climate change—"a great drying"—and the wetlands that had fostered the luxuriant growth of the Coal Age forests began to disappear. The year was now divided between a dry season, which was followed by so-called megamonsoons in regular sequence. Because of the prolonged dry spells, the moisture-loving giant club mosses

The great tides of the Bay of Fundy continue to erode its fossil-rich cliffs, opening a window in time.

and horsetails were replaced by conifers, which had previously been restricted to the highlands, as well as by gymnosperms like cycads, ginkgoes, and seed ferns.

The dominant land animals were mammal-like reptiles such as the pelycosaurs, or sail-reptiles, which have been found in the redbeds of Prince Edward Island. They sported a distinctive fanlike sail on their backs, formed by elongated neural spines of the vertebrae covered by skin. Richly supplied with blood vessels, this sail could be deployed to gather heat or to dissipate it, depending on how the animal oriented itself to the sun.

DIMETRODON

The fusion of the continents created huge deserts in the interior of Pangaea, which have been invoked to partially explain the "great dying" that brought the Permian to a resounding close, 245 million years ago, when nearly 75 percent of all amphibian and reptile families and fully half of all marine families were wiped out. Where mountains formed as the continental plates jammed together, cold temperatures, accompanied by glacial activity, may have prevailed. The productive shallow coastal seas were also reduced by the coming together of the continents. Chemical changes to the atmosphere seem to have occurred around the Permian-Triassic boundary, with dramatic changes in oxygen and carbon levels before and after, indicating a dramatic rise in global temperature followed by a precipitous drop. But taken together, these factors do not fully explain the magnitude of the greatest extinction event in fossil history.

This great dying was a dividing line in the classification of life on Earth. It was the end of the Paleozoic, or "the era of old life," and the beginning of the Mesozoic, or "the era of middle animals." The Mesozoic itself was divided into three major periods—Triassic, Jurassic, and Cretaceous—encompassing the rise, flourishing, and ultimate fall of the dominant reptile group known as the dinosaurs.

*previous spread*: The sandstone cliffs of Prince Edward Island are a vivid reminder of the desert conditions that prevailed during the Permian Age.

THIS PERIOD OF change in life on Earth was marked geologically by the breakup of Pangaea and the creation of the present-day Atlantic Ocean. In the Late Triassic (225 million years ago), the Earth's crust weakened and the continental plates began to pull apart, rupturing along fault lines roughly parallel to the present-day continental margin of the eastern seaboard. Rift valleys, similar to East Africa's Great Rift Valley, formed as blocks subsided along fault lines from Nova Scotia to South Carolina. These valleys became sedimentary basins (the so-called Newark supergroup), which began to fill with material being eroded from the now-ancient Appalachian mountain chain.

Throughout the Triassic (248–200 million years ago), sediments continued to pour into these basins. Today, two of the northerly basins are exposed in the Connecticut River Valley of Connecticut and Massachusetts and the Bay of Fundy region. Rains of monsoon intensity swelled rivers that carried coarse sediments from the Nova Scotia highlands into the Fundy Basin and lakes occupied the basin floor and finer bottom sediments piled up to impressive thicknesses—1,000 meters (3,300 feet) in places, today preserved as red sandstones brilliantly visible along the shores of the Bay of Fundy.

As the Jurassic Age succeeded the Triassic, 200 million years ago, the continental rifting process became more intense, further stressing the already weakened crust. Molten magma spewed from fissures in fiery lava fountains and spread over the red sandstones to form Nova Scotia's North Mountain, which stands guard over the south shore of the bay. This volcanic activity accompanying the breakup of Pangaea was perhaps the most widespread that has ever occurred. Volcanic rocks have been found on the margins of Pangaea from Brazil to France and from the Bay of Fundy to Morocco.

It was during this tumultuous time in Earth's history that dinosaurs emerged as the dominant animal group, and the rift valleys of the eastern seaboard contain a unique record of the very dawn of these "terrible lizards." It is unusual to find trackways and bones together, simply because the conditions for their preservation usually differ. In the Fundy Basin, however, there are abundant bones to match the trackways. Together, they tell a story of the

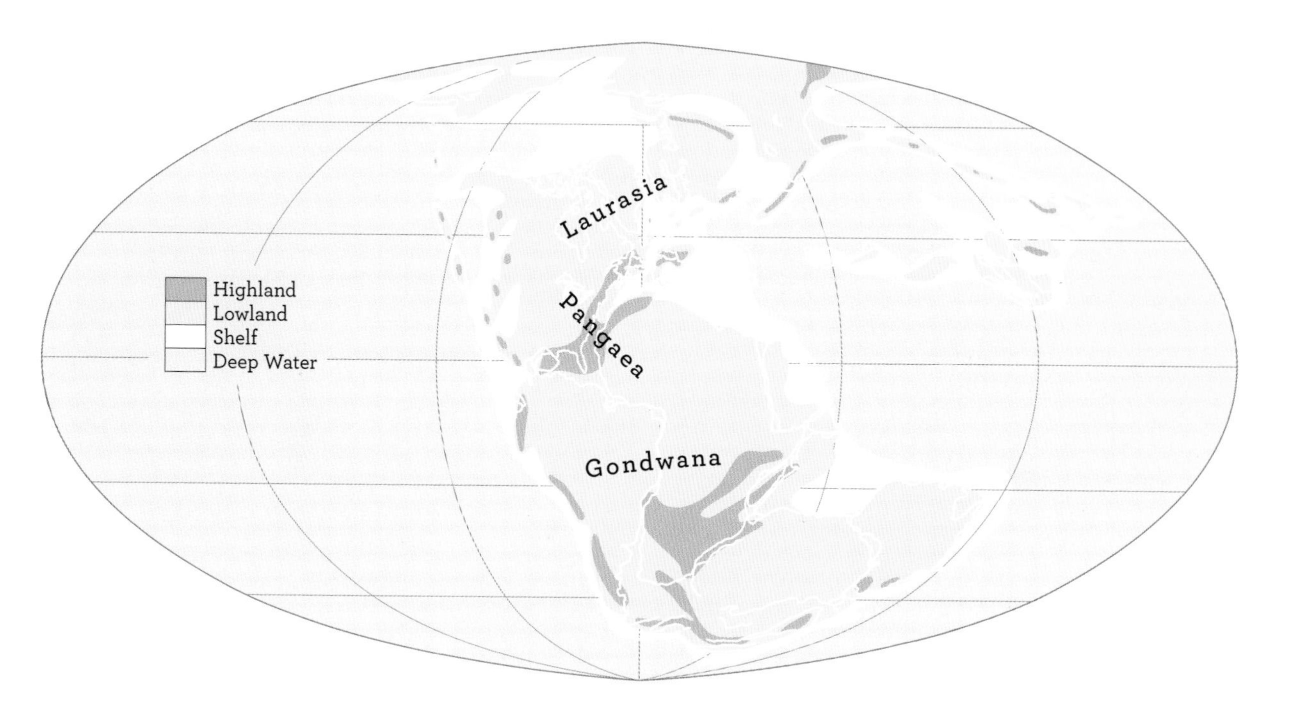


MAP: The super-continent Pangaea began to break up in the late Triassic, 225 million years ago, giving birth to the Atlantic.

demise of one group of reptiles and the takeover by another, due to yet another extinction event at the Triassic-Jurassic boundary 200 million years ago.

In the Early Triassic, the mammal-like reptiles still held sway, but more advanced types had replaced their primitive cousins, the pelycosaurs, as the dominant land animals. These were the dicynodonts ("two dog teeth") and cynodonts ("dog teeth"). Bulky plant eaters, ranging from pigsize to rhinosize, dicynodonts have been colorfully described by fossil trackway expert Paul Olsen of Columbia University as "Volkswagen Beetles with tusks." Many cynodonts were sabre-toothed carnivores that preyed on their dicynodont cousins. Their dominance would be short-lived, however.

At the beginning of the Late Triassic, there was a proliferation of a spectacular reptile group called archosaurs, or "ruling reptiles." They have best been described as "gatorlizards," to use the term coined by Canadian paleontologist Dale Russell. They included large aquatic reptiles called phytosaurs, which are considered the original crocodiles and which looked very similar to the gavial crocodiles of India, with their long narrow snouts ideally suited for snatching fish. The aetosaurs were a heavily armoured, reptilian version of the armadillo with a piglike snout for grubbing for larvae and other invertebrates. The most formidable archosaurs were the rauisuchids. These hulking quadrupeds could stand up on their hind legs and had a fearsome row of teeth comparable to those of *Tyrannosaurus*, undoubtedly qualifying them as the tyrants of the Late Triassic.

PLINTHOGOM

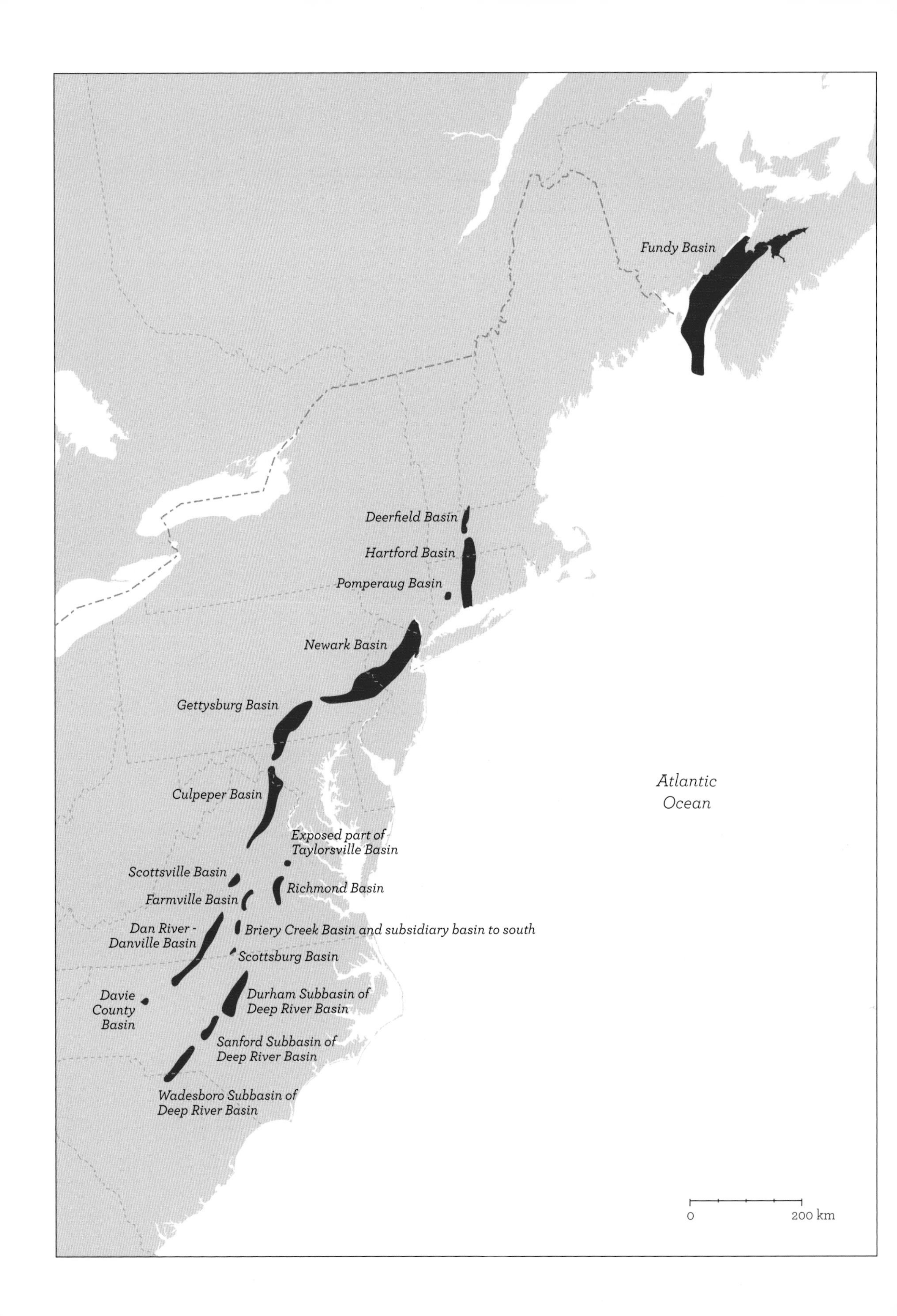

Fundy Basin
Deerfield Basin
Hartford Basin
Pomperaug Basin
Newark Basin
Gettysburg Basin
Culpeper Basin
Atlantic
Ocean
Exposed part of
Taylorsville Basin
Scottsville Basin
Richmond Basin
Farmville Basin
Dan River -
Danville Basin
Briery Creek Basin and subsidiary basin to south
Scottsburg Basin
Davie
County
Basin
Durham Subbasin of
Deep River Basin
Sanford Subbasin of
Deep River Basin
Wadesboro Subbasin of
Deep River Basin
0
200 km

Despite their formidable arsenals, rauisuchids, along with phytosaurs and aetosaurs, disappeared by the end of the Triassic and were replaced by the early dinosaurs, including a birdlike species that left footprints in the Fundy sandstones—possibly the carnivorous theropod *Coelophysis*, a member of the coelurosaurs, or hollow-boned reptiles. These were slender, fleet-footed bipeds with pointed heads, well armed with teeth serrated like steak knives. They probably hunted in packs. But there were also larger tracks made by plant-eating prosauropods, long-necked dinosaurs with small heads, which were the forerunners of the giant sauropods such as *Brontosaurus* of the Late Jurassic, the largest land animal that ever trod the Earth.

*COELOPHYSIS*

The dinosaurs were distinguished from their archosaurian ancestors by their fully upright gait, a refinement in posture that set them apart from all other reptiles. The earliest dinosaurs were, in fact, bipedal, walking on their hind legs as humans do. The bipedal locomotion resulted from certain skeletal changes. Like mammals, early dinosaurs had limbs that supported the body from beneath, rather than from the sides as in early reptiles, which were awkward sprawlers. Their erect posture gave the dinosaurs distinct advantages: they could run with greater ease and speed, a vital skill when pursuing prey or fleeing predators. Whether this adaptation was enough to allow them to outcompete the larger, more fearsome rauisuchids, for instance, remains an open question. It may be that they were simply the beneficiaries when an asteroid crashed into the Earth, causing the extinction of their larger competitors, just as an asteroid impact would cause the demise of all dinosaurs 150 million years later.

*facing page*:
MAP: As the Atlantic Ocean opened, rift valleys, or basins, formed along the margins of the eastern seaboard and today preserve a remarkable fossil record of the rise of the dinosaurs.

In the Fundy Basin the timing of the extinction can be dated with unusual precision, within 250,000 to a million years after the event, which is "the day after" in geological terms. Even though hundreds of thousands of bones and footprints have been uncovered in the Jurassic sediments just below the North Mountain basalts, there are no signs of typical Triassic animals, such as the cynodonts, dicynodonts, and archosaurs. Neither is their evidence for the emergence of new groups of animals. There are only survivors, a so-called day-after community, which is what you would expect in the case of a catastrophic extinction event such as the one that occurred at the Cretaceous-Tertiary boundary 65 million years ago, when a meteorite slammed into the Earth, wiping out the dinosaurs and leaving behind the Chicxulub Crater in the Gulf of Mexico and a telltale iridium layer at the Triassic-Jurassic (T-J) boundary.

Although no crater or iridium layer has been discovered to date at the T-J boundary, other lines of evidence point to the probability that an asteroid or

## QUARRYING FOR TRACKWAYS

THE FIRST evidence of dinosaurs on the North American continent was discovered in 1802 by a young farmboy Pliny Moody. While plowing his father's field in South Hadley, Massachusetts, he uncovered a red sandstone slab containing three-toed dinosaur footprints. Pliny and his neighbors jokingly attributed the trackways, which were proudly displayed above a door in the Moody farmhouse, to "Noah's raven," thinking they had been made by ancient birds.

Subsequently, many dinosaur trackways came to light in the New England quarries that supplied brownstone for Manhattan's mansions. Reverend Edward Hitchcock, the president of Amherst College, in Massachusetts, spent a lifetime collecting and writing about the Connecticut Valley trackways. He died in 1864, persisting in his view that the tracks were made by birds: "Now I have seen, in scientific vision, a [wingless] bird, some twelve or fifteen feet high,—nay, large flocks of them,—walking over the muddy surface, followed by many others of analogous character, but of smaller size."

We now know that Hitchcock was essentially correct in his analysis, for modern birds have since been shown to be the direct descendants of dinosaurs. We also know that many of the early dinosaurs were the size of chickens and turkeys, or even smaller. Tracks made by a dinosaur no bigger than a robin were discovered in the Bay of Fundy in 1984.

One of the most important repositories in the United States is Riker Hill Fossil Site (also referred to as Walter Kidde Dinosaur Park) in Roseland, New Jersey. This 6.4-hectare (16-acre) site preserves a bounty of dinosaur tracks. Three teenagers, Paul E. Olsen, Tony Lessa, and Bruce Lordi, began visiting the old quarry in 1968, uncovering more than a thousand dinosaur tracks. They sent a cast of one of the largest—*Eubrontes gigantes*—to President Nixon in hopes of thwarting a development plan for the site, which was declared a National Natural Landmark in 1971. Olsen, currently a professor at Lamont-Doherty Earth Observatory of Columbia University, went on to make major dinosaur discoveries in the Fundy Basin and to become an authority on the fossil trackways of eastern North America.

The Bay of Fundy shoreline (left) holds a unique record of an extinction event that marked the Triassic-Jurassic boundary, roughly 200 million years ago; ferns (right) are the first plants to colonize a disturbed environment.

meteorite impact accounted for the mass extinction, which saw the demise of nearly half of all creatures on Earth. In the Fundy Basin as well as in New Jersey there is a very abrupt changeover in vegetation from a high-diversity Triassic plant community to a low-diversity Jurassic community, the kind of floral change that is consistent with an asteroid impact. Furthermore, in New Jersey, a fern spike occurs, where the plant community is suddenly 90 percent ferns. This spike is sandwiched exactly between the Triassic-type and Jurassic-type plant communities, indicating that something unusual happened. The sudden upswing in the proportion of ferns is similar to what has been observed at the Cretaceous-Tertiary boundary. Ferns are generally first colonizers of a disturbed environment. When the island of Surtsey was born off Iceland in 1963, for example, ferns were the first plants to grow there, and when Krakatoa exploded in the 19th century, ferns were the first plants to colonize its shores. Although the abrupt change in plant communities and the faunal evidence of a day-after community constitute circumstantial evidence, they point a finger at a catastrophic impact as the cause of extinction.

Along with the first dinosaurs, the early crocodilians, pterosaurs, turtles, and mammals survived this extinction. In the Jurassic, this assemblage of survivors constituted the modern world in embryo. For the next 100 million years, however, dinosaurs dominated the Earth's biota, and our own lineage, the mammals, survived in the shadows.

The supercontinent Pangaea began to break apart in the Jurassic, causing the massive volcanism that created the North Mountain in Nova Scotia as the crust weakened and magma welled up. Even though the continental plates began to pull apart, the fossils in the Fundy Basin indicate that in earliest Jurassic times the landmasses remained connected. For example, the sphenodontids found in Fundy, sturdily built little reptiles with parrotlike beaks, are identical to fossils from England, suggesting that, 200 million years ago, the Atlantic Ocean had not yet sufficiently opened to isolate these two populations. Similarly, the long-legged crocodilian *Protosuchus* from Fundy is identical to specimens from China and South America, strongly suggesting that the continents still formed a single landmass at least into the earliest Jurassic.

*Ice and the Modern Coast*

But the modern-day Atlantic Ocean was inexorably being born. As Eastern Canada drifted away from what is now southern Europe and northwest Africa, parts of those now-distant continents remained welded to eastern North America (such fragments of crustal material are called terranes). The pieces left behind form the Avalon Terrane and today include eastern Newfoundland and parts of Nova Scotia and Maine. Other remnants of Avalonia can be seen in Ireland, England, continental Europe, and North Africa.

Throughout the Jurassic and Cretaceous, the Appalachians continued to be worn down, and the sediments were deposited as a coastal plain in the Gulf of Maine and on the Scotian Shelf. The sea level rose and fell during this period, so the seashore was sometimes 100 kilometers (60 miles) or more south of Nova Scotia and at other times only a few kilometers from the current coastline. The erosion of the Appalachians created a shallow, seaward-sloping shelf like that found today south of Long Island.

There are very few exposed rocks along the contemporary Atlantic coast from the Tertiary period, 65 to 2 million years ago, that followed the demise of the dinosaurs at the end of the Cretaceous. The disappearance of the dinosaurs paved the way for the rise of the mammals, and so the Tertiary became known as the Age of Mammals. During this period, parts of the Appalachians were uplifted, including western Newfoundland and the Cape Breton Highlands; at the same time, the southern Gulf of St. Lawrence emerged above sea level.

The climate cooled as the North American Plate migrated slowly northward and as the tropical Tethys Ocean, which wrapped around the world at the equator during the Cretaceous, began to close. The last equatorial "gateway" was slammed shut with the development of the Isthmus of Panama 3 million years ago. This obstruction channeled the Gulf Stream farther northward. The warm water, in turn, supplied the necessary moisture for the production of snow in northwestern Europe and northeastern North America and, critically, the development of the Arctic ice cap and formation of cold currents around the North Pole. This cooling set the stage for the onset of glaciations in the Quaternary period, which is divided into the Pleistocene epoch, 2.5 million to 12,000 years ago, and the Holocene epoch, the interglacial in which we are now living.

Glaciers are moving bodies of ice and snow. They form when winter accumulation is greater than summer melting and the snow begins to build up. As it becomes thicker, the weight of the snow causes ice to form at its base. When a thickness of about 60 meters (200 feet) is reached, the glacier begins to flow at the rate of a few meters per year.

Ice Ages, such as the Pleistocene, are characterized by climatic cycles of about 100,000 years each during which the ice sheets alternately advance and retreat. During these cycles, glacials last 60,000 to 90,000 years and interglacials 10,000 to 40,000 years. These cycles are known as Milankovitch cycles, for the Serbian mathematician who first calculated them. The rhythm of these cold periods relates to three planetary cycles that affect the distribution of solar radiation between the Southern and Northern Hemispheres.

Glaciers, such as this one clinging to the Labrador coast, shaped much of the contemporary Atlantic coastline.

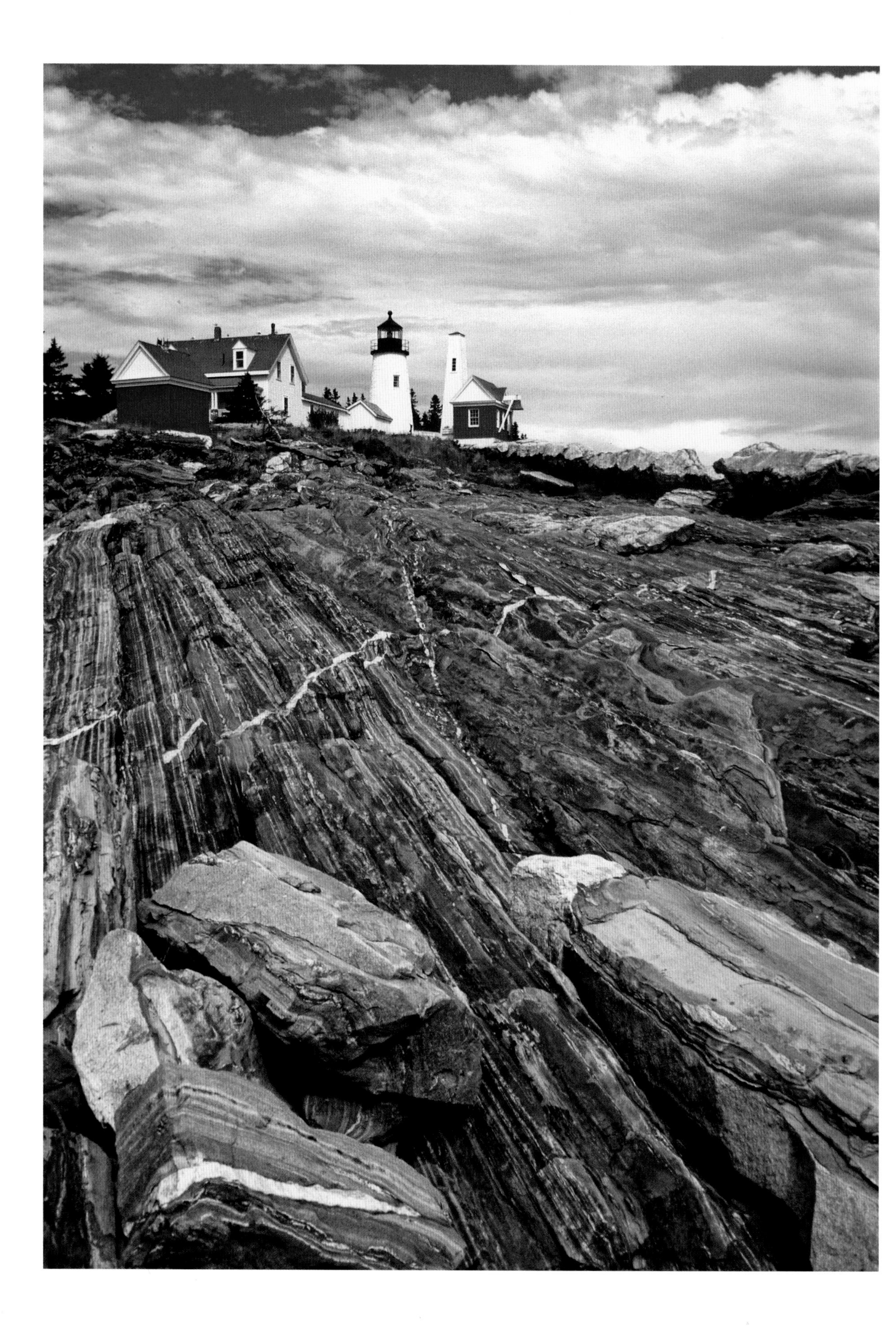

The Northern Hemisphere currently receives less solar energy than its southern counterpart, periodically causing the ice sheets to grow. The largest was the Laurentide ice sheet, which originated in the Laurentian region of Canada and once covered all of North America east of the Rockies, reaching as far south as St. Louis, Missouri. There were four major advances of the Laurentide ice sheet. The last was the Wisconsinan ice sheet, which gouged out the Laurentian Channel in the Gulf of St. Lawrence, eventually filled the gulf and the Bay of Fundy and advanced across Nova Scotian to the edge of the Scotian Shelf. Twenty-five thousand years ago, it plowed across New England as far south as Long Island, but by eighteen thousand years ago a warming of the global climate had caused it to retreat north of Cape Cod.

The Wisconsinan ice sheet has had profound effects on the shaping of the landscape and the life found along the Atlantic coast today. Although the Atlantic coast from Labrador to Florida shares a similar geological history, largely the consequence of plate tectonics and the mountain-building episodes that accompanied continental drift, it is the advance and retreat of the great ice sheet that is responsible for the different character of the northern and southern sections of the contemporary coastline.

*facing page*:
The grinding advance and retreat of glaciers left their dramatic marks on the coastline of Maine.

Sediments that have eroded from the Appalachians form the Atlantic Coastal Plain, an apron of sand and mud that lies between the mountains and the ocean and extends 3,500 kilometers (2,200 miles) from the New York Bight southward to Florida. It slopes gently, at an angle of about one degree, toward the Atlantic, then continues under its waters to form the continental shelf. By contrast, the coastal plain is generally missing in the north, where the glacier acted like a great bulldozer blade, stripping most of the coastal plain sediments north of Long Island and depositing them in the south. Therefore, north of Long Island along the coast of New England and in Maritime Canada the coast is largely rocky and very irregular, whereas south of Long Island it is quite linear and regular and is protected by chains of barrier islands, which were formed from the reworking of those sediments by wind and currents.

Cape Cod, Nantucket, and Martha's Vineyard are part of the Atlantic Coastal Plain and more closely resemble the Atlantic coast of New Jersey than the rockbound coast of New England north of Boston. They are characterized by low relief and reflect their shaping by glacial action. The sea continues to rework this material to create barrier spits, and the wind also acts as a sculptor, molding impressive dunes.

At its maximum the ice was probably 500 meters (1,600 feet) thick over Cape Cod, whereas in northern New England it was 1,500 meters (5,000 feet) in thickness, covering the White Mountains of New Hampshire. The front of the Wisconsinan ice sheet was not straight but lobate. In the vicinity of Cape Cod, there were three major lobes: one occupying the Great South Channel to the east of the cape, a middle one in Nantucket Sound and Cape Cod Bay, and the westerly one in Buzzards Bay, to the west of Martha's Vineyard. The triangular shapes of Nantucket and Martha's Vineyard can be traced to the fact that they were formed in the angle between two of the glacier's lobes. Similarly, the bent-arm shape of Cape Cod developed between glacial lobes.

The cape and its islands are formed of a patchwork of glacial deposits. As they make their grinding advance, glaciers erode underlying material and incorporate it into the base of the ice—a mixture of everything from boulder-sized rocks to microscopic clay particles. Taken together, this heterogenous mixture is known as glacial till.

As the glacier begins to melt, the water flows down through crevasses and tunnels within the glacier and forms meltwater streams at the terminus of the ice. These meltwater streams usually make a braided pattern, consisting of numerous broad channels separated by gravel bars. The sediment carried beyond the glacier and deposited downstream is called outwash. Outwash plains are the most common landforms on Cape Cod and the islands, and a single outwash plain makes up most of Martha's Vineyard and upper Cape Cod.

End, or terminal, moraines are a second type of glacial deposit formed when a glacier has been parked in one place for an extended period of time. They mark the farthest extent of the glacier's advance. Broad and elongated, they are also distinguished by a series of hummocky ridges. In the cape region, they seem to have been formed when the glacier plowed up a series of thrust sheets—horizontal layers set up on end—each forming a hummock or ridge. Today it is possible to determine the extent of the Wisconsinan ice sheet by tracking these end moraines. There are two long east-trending moraine belts: one loops from western Long Island through Block Island and Martha's Vineyard to Nantucket and marks the southern limit of the Wisconsinan ice sheet's advance. A more northerly belt, formed when the ice sheet advanced a second time after initially retreating, stretches from Long Island to eastern Connecticut, southern Rhode Island, and the inner side of Cape Cod.

The retreat of the glacier was relatively fast, in geological terms. Cape Cod and islands emerged in just a few thousand years, between 23,000 and 19,000 years ago. Dramatic changes in sea level accompanied the advance and retreat of the glaciers. At the global maximum of the Ice Age, so much seawater was

Caribou followed the retreating glaciers; they were quickly followed by Holocene hunters.

tied up in the glaciers that the sea level was 120 meters (400 feet) lower than it is today, and broad expanses of the continental shelf were exposed beyond the limit of the ice sheet. Occasionally fish draggers dredge up peat, bones, and teeth as evidence of the life that once existed there. Among the more intriguing signs of this Pleistocene ecosystem are the teeth of mastodon and mammoth, which would have roamed a spruce-clad parkland that is now submerged continental shelf south of Cape Cod. To the north, Georges Bank in the Gulf of Maine was also exposed and has also yielded mastodon and mammoth teeth. Mammals more familiar to contemporary residents of the Atlantic coast, such as deer, moose, and bear, also found a niche in this Ice Age world. And northern species, such as caribou, lived in the shadow of the glacier and followed it north as it retreated, as did their human hunters.

Although the Laurentide ice sheet had retreated north of the St. Lawrence River by twelve thousand years ago, small ice caps and permanent snowfields remained in Nova Scotia, and perhaps in New Brunswick and the Gaspé, for another one thousand years, and the glacier may have advanced again during

The glaciers are gone, but wind, wave, and ice continue to sculpt the Atlantic coast.

a dramatic cooling period known as the Younger Dryas event, 12,800 to 11,500 years ago. Temperatures in the North Atlantic region dropped dramatically, by as much as 5° to 7°C (41° to 45°F), in as little as a decade. Although this striking example of sudden climate change is not fully understood, it is thought to have been due in part to the influx of vast quantities of cold freshwater from the melting of the glaciers. The cold water temporarily put a lid on the North Atlantic thermohaline circulation, which brings warmer water into the Northern Hemisphere.

The Younger Dryas event ended approximately eleven thousand years ago, after which the glaciers disappeared from the Atlantic coast for good, ushering in the interglacial period in which we are now living—the Holocene. In the postglacial period the sea level has seesawed in response to complicated forces related to the retreat and melting of the glaciers. At first, the sea level rose rapidly as the glaciers returned to the oceans vast reservoirs of water that had been tied up in ice. These changes are referred to as "eustatic." Coastal river valleys were drowned, and in Maine the sea reached 100 kilometers

(60 miles) inland from the present-day coastline. Prince Edward Island was cut off from the mainland seven thousand years ago, and waters in the Gulf of St. Lawrence transformed the highlands into the island of Anticosti and the Magdalen Islands.

As the glaciers retreated, however, the land rebounded in response to removal of the glaciers' weight. The glaciers had depressed the land by a few hundred meters, pushing down on the crust overlying the relatively plastic asthenosphere, which was squeezed outward at the edge of the ice sheet, causing the land or seafloor to bulge upwards. As the weight of the glacier was lifted, however, the land as a whole rebounded, just as a trampoline rebounds when weight is removed. Such changes in sea level due to the ups and downs of the lithosphere, or Earth's crust, are known as "isostatic" changes. The land reached its maximum height about six thousand years ago and then slowly began to subside. At that time, Georges Bank was an island, but waters now began to flood over it, reaching into the Bay of Fundy, where they would eventually propagate the great tides there. Today, the only place in the Atlantic region of North America where the land is still rebounding from the removal of the ice burden is northern Newfoundland and Labrador. Elsewhere the land is still subsiding, resulting in a steady, if gradual, rise in sea level at an average rate of 8 centimeters (3 inches) per year.

By two thousand years ago, Cape Cod and its islands had assumed the shapes that they have today, even though the shoreline was still several kilometers from its present-day position. Great sea cliffs faced the Atlantic, with narrow beaches at their base. Barrier spits and islands formed by longshore currents protected the bays and inlets, and the salt marshes sheltered in their lee. Essentially, the coastline had assumed its contemporary character.

*The Atlantic Coastal Plain*

Today, the Appalachians serve as a rugged backbone of eastern North America, with the land sloping away on either side—in the west to the Mississippi and in the east toward the Atlantic Ocean. Since the opening of the present-day Atlantic in the Triassic period, sediments have been washing down from these once-towering mountains and pouring into the deep rift basins that formed along the eastern coast of North America.

As we have seen, a geological divide in the character of the Atlantic coast runs in an east-west direction through Long Island, New York, which can be traced to the more recent glacial history. North of Long Island, the bedrock of the Appalachians extends to the shoreline. The coastline is irregular where it has been gouged out by the action of the glaciers, and the rocks are resistant to erosion. The rugged coast is distinguished by rocky headlands and numerous small islands and shoals. South of Long Island, all the way to Florida, the Atlantic coast is markedly different, being more linear and flanked by chains of barrier islands that run parallel to the coast. The unconsolidated sediments of the Atlantic Coastal Plain are the building materials for this nearly unbroken chain. Barrier islands usually form on trailing edge coasts, such as the eastern side of the North American continent. The broad, low relief of the coastal plain and shallow continental shelf, together with the abundant supply of sediments, provides the ideal conditions for the formation of barrier islands.

Three fundamental ideas about how barrier islands come into being have been proposed and continue to be debated and elaborated on. According to a theory put forward in the 1960s by the late John Hoyt, of the University of Georgia, barrier islands are formed from the drowning of coastal ridges, and this theory seems to apply to the majority of such islands. A second theory holds that they begin as spits created by longshore drift, which are later breached, disconnecting them from the mainland. The third theory posits that barrier islands may form as a result of the emergence and upward-shoaling of offshore sandbars. Each of these mechanisms applies in different cases but all, to varying degrees, are tied to the changes in sea level caused by the waxing and waning of the glaciers.

The sea level rose very rapidly between eighteen thousand and fifteen thousand years ago as the icecaps gave up their water—a rate of 1 meter (3 feet) per century, or over 100 meters (300 feet) during this initial period of glacial melting. As the sea level rose, the ridges of mainland beach dunes were breached, the lowlands behind these ridges were flooded to form lagoons, and the ridges became barrier islands. The Hoyt theory of barrier island formation seems to apply to most barrier islands along the southeast Atlantic and Gulf of Mexico coasts, including the Outer Banks of North Carolina.

Along the northeast coast, from New York northward in the region directly affected by the glaciers, barrier islands seem to be most commonly formed when spits created by longshore drift are later breached. First, barrier beaches are formed by the erosion of a sea cliff or associated beach. Longshore currents then transport the sand parallel to the shore to form a spit. This elongated, fingerlike ridge is attached at one end to the mainland but curves into open water. At some point, the spit is breached to create an island. The barrier island system along the south shore of Long Island evolved in this way, as did Monomoy Island of Cape Cod.

Barrier islands, like these on the north shore of Prince Edward Island, have been formed from the sediments washed down from the ancient Appalachian Mountains and form an almost continuous chain from Long Island to Florida.

Barrier islands can also develop from submarine bars. As waves transport material landward, an offshore bar is created that eventually becomes an emergent island when it builds above sea level. It seems likely that some small barrier islands along the Gulf Coast originated in this way.

However they are formed, barrier islands are dynamic entities that migrate and change shape almost like living things. Storms strip away sand from the upper beach and dunes and carry it toward the sea, where it is stored temporarily offshore as a storm bar. Then, in the following weeks, this sand borrowed from the beach is gradually returned and again "welded" to the beach. This system of sand-sharing reaches a state of dynamic equilibrium, in which the sands are subtracted from and then added back to the beach. However, severe storms sometimes breach the dune or push through low areas in the dune ridges. When they lose energy during this landward surge, they deposit the sand as an overwash fan. This overwash process reshapes barrier islands over time, and eventually the whole island migrates landward. As the island moves, however, the sequence of habitats re-establishes itself. A wave-swept sandy beach faces the sea, sand dunes rear up in the middle of the island, and a lagoon fringed with salt marshes forms between the island and the mainland.

The creation of barrier islands is a striking example of geology in action, of the dynamic processes that constantly reshape the coastline, just as the mostly hidden process of plate tectonics is moving continents and in the process making mountains and opening oceans.

# 3

# THE ATLANTIC HINTERLAND

## *Forests of the Atlantic Coast*

This is the forest primeval. The murmuring pines and the hemlocks,
Bearded with moss, and in garments green, indistinct in the twilight...
Loud from its rocky caverns, the deep-voiced neighboring ocean
Speaks, and in accents disconsolate answers the wail of the forest.

—*Evangeline, A Tale of Acadie*, HENRY WADSWORTH LONGFELLOW

THESE FAMOUS OPENING lines of one of the best-known and most beloved American poems of the 19th century allude to a forest that today we would describe as old growth. The clearing of land for agriculture, intensive logging, and epidemic fire and pestilence followed closely upon European colonization of eastern North America, and only a few small patches of these ancient trees, "bearded with moss," can be found today. It is estimated that, when Europeans arrived, two-thirds of the continent was covered with old-growth forests. But by the middle of the 19th century, crops and pastures covered nearly three-quarters of the arable land in southern and central New England. Although today the proportions are exactly reversed, since farmland has been abandoned in the last two centuries, the forests are younger and in most instances composed of different proportions of species than the "forest primeval."

Two major biomes—large regional groupings of plants that

< Mixed forests clothe the eroded peaks of the Appalachian Mountains.

have the same climax vegetation and share a similar environment—meet in the area centered on the Gulf of Maine. Here the temperate broadleaf and mixed forest, which extends from the Great Plains to the Atlantic coast, and from the Gulf of Mexico to Maine, meets the boreal forest, which sweeps across the northern half of the continent. The former is characterized largely by deciduous trees, which shed their leaves in the fall, and the latter by conifers, or evergreens.

The boreal forest is largely dominated by the spruces (white, black, and red) and balsam fir, which are tolerant of short growing seasons and are hardy enough to survive temperatures that can dip to -40°C (-40°F). Many of these evergreens can also tolerate nutrient-poor and waterlogged soils. To the south is a primarily deciduous forest; in between, in Maritime Canada and northern New England, there is a transition zone characterized by a mixture of coniferous and deciduous species dependent on variations in the local elevation, soil type, and climatic conditions. This is the New England/Acadian Forest type alluded to in Longfellow's poem.

*facing page*:
MAP: Terrestrial ecoregions of the Atlantic coast.

*New England/Acadian Forest*

The New England/Acadian Forest is the largest ecoregion bordering the Atlantic coast within the temperate broadleaf and mixed forest biome, and despite four hundred years of alteration in the wake of European settlement, it is also the most intact.

Although this New England/Acadian Forest region includes many thousands of kilometers of coastline, the region does not have a typical maritime climate, largely because the prevailing winds are westerly, blowing off the continent. Even so, along the Gulf of St. Lawrence, the Bay of Fundy, and Nova Scotia's Atlantic shore, the climate is somewhat moderated by the closeness of the Atlantic Ocean, where the Gulf Stream exerts its moist, warming influence, resulting in milder winters and cooler summers than in the center of the continent. Because the entire region is peninsular, the influence of the ocean is felt for 200 kilometers (125 miles) inland.

Historically, the New England/Acadian Forest was dominated by shade-tolerant, long-lived species such as sugar maple, beech, eastern hemlock, and red spruce, with a significant amount of white pine and yellow birch mixed in. Perhaps the defining characteristic of the New England/Acadian Forest is the dominance of red spruce. Pollen evidence taken from lakes in northern Maine, Nova Scotia, and New Brunswick, and from bogs in Prince Edward Island, indicate that the typical New England/Acadian Forest tree species

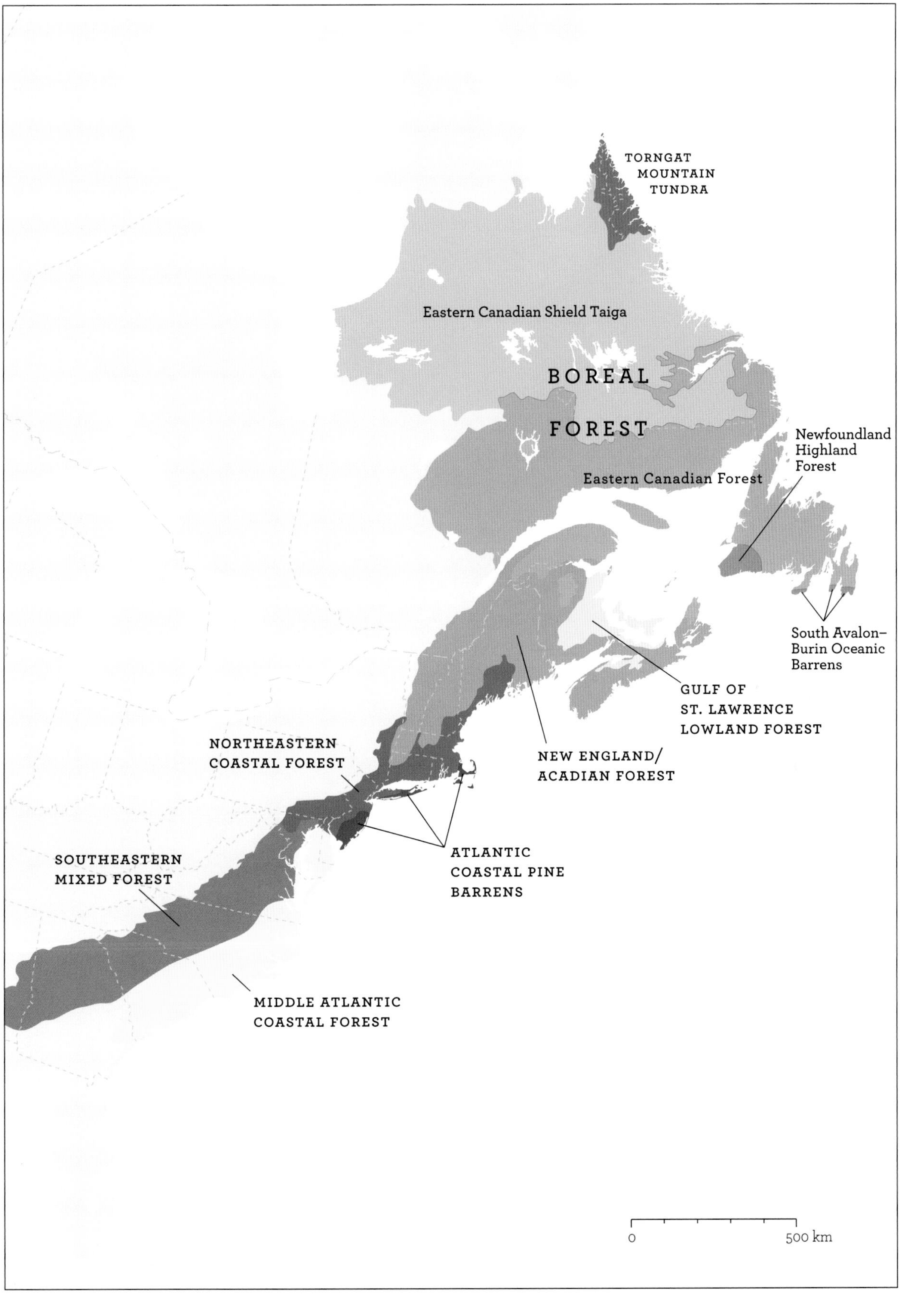

TORNGAT MOUNTAIN TUNDRA

Eastern Canadian Shield Taiga

BOREAL FOREST

Eastern Canadian Forest

Newfoundland Highland Forest

South Avalon–Burin Oceanic Barrens

GULF OF ST. LAWRENCE LOWLAND FOREST

NEW ENGLAND/ ACADIAN FOREST

NORTHEASTERN COASTAL FOREST

ATLANTIC COASTAL PINE BARRENS

SOUTHEASTERN MIXED FOREST

MIDDLE ATLANTIC COASTAL FOREST

0

500 km

A yellow birch, a climax species of the New England/Acadian Forest, clings to a rocky outcrop.

have been growing in these areas for at least five thousand years. At the time of European settlement, 60 to 85 percent of northeastern North America was covered by old-growth trees, having an average age of greater than one hundred years, whereas today less than 1 percent of Nova Scotia's forests, for example, is considered to be old growth.

Nicolas Denys, the author of the first natural history of Acadia (the old French territory that encompassed all three Maritime Provinces and parts of Maine), commented on the bountiful forests. The trees were so large that they created a kind of wilderness parkland through which it was possible, he claimed, to pursue a moose on horseback. Other early chroniclers commented on the great size of the trees—yellow birch 3 to 4 meters (10 to 13 feet) in circumference, beech 2 meters (6.5 feet) in diameter, and white pine 38 to 53 meters (125 to 175 feet) tall.

Before European settlement, these late-successional forests, consisting of climax species, were rarely subject to catastrophic disturbances, such as fire caused by lightning. In New Brunswick and Nova Scotia, only about 1 percent of fires are caused by lightning today, with the remainder attributed to human causes, and charcoal evidence indicates that large-scale fires occurred only every eight hundred to a thousand years in northeastern Maine and New Brunswick. The long fire-cycle intervals meant that there was enough time between disturbances for the forest to develop into an old-growth, late-successional type—a climax forest.

The climax species—sugar maple, yellow birch, American beech, red spruce, eastern hemlock, and eastern white pine—live an average of three hundred to four hundred years, and their natural regeneration occurs within canopy gaps that result from small-scale disturbances, such as blowdown of older trees, rather than stand-replacing events, such as the large fires typical of boreal forests. The result is a self-sustaining forest with trees of various ages, a multistoried canopy, and some dead or dying trees in various stages of decay, as well as large rotting logs on the ground. Over time, this forest reaches a steady state, in which composition and age structure change little over time.

Although this description applies to the classic old-growth New England/Acadian Forest type, once common to much of Maritime Canada and northern Maine, the situation on the ground is often different. Dominant species or species associations vary, depending on local climatic or soil conditions and history of exploitation. In the moist coastal climates of Nova Scotia and southeastern New Brunswick, for example, red spruce occurs with early-successional species such as black spruce, red maple, trembling aspen, and white birch. These maritime forests, like those found in the adjacent New England states, were probably once dominated by red spruce.

The Atlantic coast of Nova Scotia has a boreal coastal forest, dominated by early successional forest types as a result of its long history of human occupation and fire disturbance. In this windswept region, white spruce is the dominant conifer species, since it is tolerant of salt spray. The stunted dense forest typical of this region has a mixture of conifers, including white and black spruce and balsam fir, with a few red maple and white birch mixed in. A more diverse and robust mixed-wood forest—the Acadian-boreal coastal type—occurs in the protected coves along the Atlantic coast of Nova Scotia, as well as along the entire coastline of the Bay of Fundy and Northumberland Strait. The coastal region surrounding the highlands of northern Cape Breton produces a unique boreal transition forest in which long-lived yellow birch occupies the upper story, and short-lived but shade-tolerant balsam fir makes up the understory.

Boreal forests occur at higher elevations of the Maritimes Highlands, namely in the Cape Breton Highlands and in the northwestern corner of

New Brunswick. Being surrounded on three sides by the ocean, the Cape Breton Highlands have a more maritime climate and are dominated by balsam fir, which is less susceptible to fire but falls victim to spruce budworm infestations and massive blowdowns on a sixty-to-eighty-year cycle. Other minor components of this forest include white birch, American mountain ash, red maple, white spruce, and black spruce.

On lower elevations and in the inaccessible river canyons of the highlands, it is still possible to find a mixed, shade-tolerant deciduous forest of sugar maple and yellow birch, which are among the last vestiges of old-growth forest not only in Nova Scotia but also in northeastern North America. They represent rare examples of a virgin ecosystem, largely unaffected by the depredations of human activity over the four centuries since the settlement of the region.

RED SPRUCE

The dry coniferous forests of the northern New Brunswick highlands have been battered by repeated fires, logging, and insect attacks, especially by spruce budworm. The result is a forest dominated by balsam fir, white birch, aspen, and white and black spruce, with a minor mix of both deciduous and conifer species, including red and sugar maple, yellow birch, red spruce, and eastern white pine.

It is possible to trace a historical timeline of human activities that have transformed the forest in Maritime Canada and northern New England. Shipbuilding was one of the first industries and had perhaps the longest-lasting effects on the composition and quality of the New England/Acadian Forest. For a century, beginning in the 1760s, this coastal enterprise employed the best trees of the old-growth forests: birch, ash, beech, maple, tamarack (or larch), cedar, oak, and pine. The tallest and straightest white pines were marked with the King's Broad Arrow and shipped off to England to rig the British naval vessels. Between 1820 and 1840, pine along all the rivers large enough to float logs were also cut into square timber for export to Britain. When the tamarack, yellow birch, and beech were depleted, shipbuilders turned to spruce.

Beginning in the late 18th century, hemlock bark was used in the tannery industry, a practice that continued until the early 20th century. Cedar, prized for its durability, was removed for fence posts and shingles. Over time, high-grading—the removal of the biggest and best trees—of the mixed-wood forests converted them to hardwood, usually with a large balsam fir component. The pulp and paper industry started up in the 1920s and targeted black spruce, white spruce, and jack pine.

Ice and snow illuminate coniferous forests that occur at higher elevations in New England and Maritime Canada.

Since the 1950s, clear-cutting has been the harvest method of choice and along with the use of herbicides has led to the conversion of mixed-wood stands to even-aged, single-species softwood plantations. Agriculture also removed much of the original forest, and fire, frequently used for land clearing, caused much inadvertent loss and degradation of the forest. The Great Miramichi Fire of 1825 resulted from the joining of many separate land-clearing fires after a summer drought and destroyed roughly 20,000 square kilometers (7,700 square miles) of forest land, making it the largest single fire recorded in North America.

The increase in balsam fir, favored for a number of human activities, has fueled the severity and frequency of outbreaks of spruce budworm (an insect endemic to the region) in the 20th century. Human-introduced diseases have also had a major impact on hardwood species. Beech bark disease, which arrived through the port of Halifax in the late 19th century, has decimated what was once among the most common hardwood species in the region. The loss of this dominant hardwood has significantly affected the ecology of maritime forests. Dutch elm disease, introduced into Ohio from Europe around 1930, has also ravaged this once-magnificent bottomland hardwood, with ecological as well as aesthetic consequences.

THE MOST COMPREHENSIVE study of the changes wrought to the original New England/Acadian Forest in New Brunswick was carried out in Kouchibouguac National Park and the surrounding area of southeastern New Brunswick, bordering the Gulf of St. Lawrence. (This region is sometimes set apart from the New England/Acadian Forest as the Gulf of St. Lawrence lowland forest, where the maritime climate encourages the growth of hardwoods, which are mixed with red spruce, balsam fir, eastern hemlock, and white pine.) Researchers used four lines of evidence to determine the makeup of the forest two hundred years ago: written documents of the early travelers, explorers and settlers; records of "witness trees," used to mark corners of original land grants; timber petition records from 1820 to 1840, the peak time for harvesting of square timber for export to Great Britain; and finally, ecosystem archaeology, which determines what trees had grown in areas cleared for farming by fire, by examining charcoal and other identifiable macrofossils.

The last two hundred years have seen a dramatic shift from mostly late-successional, shade-tolerant species such as hemlock, cedar, and yellow birch, to early successional species, including black spruce, jack pine, and aspen. The forests are less diverse, with six tree species constituting 95 percent of the contemporary forest, compared with nine species two centuries ago. White pine has declined, most of it having been removed from riparian zones between 1820 and 1840. It has been replaced by jack pine, a product of the frequent fires in the settled period. (The study showed that before European settlement the interval between major fires was 2,900 years.) Finally, the dominant, long-lived, eastern hemlock—the bearded one of Longfellow's poem—has been replaced by short-lived species such as poplar and balsam fir. Poplar has increased by some 800 percent, while beech has drastically declined. In short, the forest today is much younger and less diverse and, in composition, more like the boreal forest to the north.

*Forest Denizens, Old and New*

Over the entire New England/Acadian Forest region, the forest has become simplified, with a loss of ecosystem diversity. Inevitably, this has had a negative effect on a variety of species that were formerly adapted to old-growth

## PRECOCIOUS CONIFERS AND DOMINANT DECIDUOUS TREES

The conifers, or evergreens, can thrive under the harshest conditions, but colorful deciduous forests now dominate the woods of eastern North America.

CONIFERS WERE the first trees to evolve, 250 to 265 million years ago, from ancient gymnosperms—nonflowering naked seed plants. These upland plants descended into the lowlands as the great swamplands, lakes, and flood plains of the Coal Age began to dry out in the Permian. Each tree usually contains male and female cones. The male cones release millions of spores to fertilize the female cone, a reproductive strategy that harkens back to marine plants. The wind-borne spores are released in the spring, and the seeds mature in the fall, usually in the same year in which pollination occurs, though pine seeds mature after two growing seasons. The seeds, which have one or two wings, are dispersed on the wind and, in temperate zones, require a chilling period before germination.

The angiosperms, or flowering plants, arose in the mid-Cretaceous, about 100 million years ago, and include the deciduous trees, whose ancestors did not appear until the end of the Cretaceous. The angiosperms proved to be a very successful group of plants, which today includes some 250,000 species, compared with 550 conifer species. Unlike gymnosperms, the angiosperms conceal their seeds inside an ovary. This strategy conferred a number of advantages, protecting them from infections, drying, and insects. By the Early Tertiary (65 to 1.8 million years ago) deciduous forests virtually covered North America, and by the Late Miocene, 15 million years ago, a number of modern tree families, such as beech and maple, had become established.

Conifers advanced and retreated with the glaciers and today are dominant at higher elevations of the Appalachians. The largely deciduous forests of eastern North America evolved as an adaptation to winter but developed where growing seasons were relatively warm, long, and humid. A number of bird and mammal families co-evolved with this highly successful plant group, including songbirds and deer, which continue to be important components of deciduous as well as coniferous forests.

Cavity-nesting species, like the hairy woodpecker (left) and boreal owl (right), are dependent upon old-growth trees.

forest. Although few animal species are critically dependent on the old-growth forests of eastern North America, a number of vertebrates exploit certain elements to complete their life cycle. The cavity nesters, such as owls, bats, and woodpeckers, require large, partially decayed older trees to provide them with nest sites. Pileated woodpeckers and barred owls are two species that require large hardwoods. But birds use different components of trees, as pointed out by Richard H. Yahner in *Eastern Deciduous Forest: Ecology and Wildlife Conservation*:

> Woodpeckers and nuthatches forage along trunks and large branches and chip away at the bark of trees in search of food; warblers and vireos forage for insects on small branches and leaves in the forest canopy; flycatchers and tanagers "sit and wait" for flying insects in open areas of the forest; wood thrushes, eastern towhees, ovenbirds, and some warblers feed on or near ground level; and hawks and vultures soar within and over forests searching for food.

Among the species that have been affected by the loss of old, shade-tolerant hardwood forest types are downy woodpecker, pileated woodpecker, eastern wood-pewee, white-breasted nuthatch, and black-throated blue warbler. Similarly, the decline of old spruce-fir forest types has reduced habitat for a number of bird species, including red crossbill, white-winged crossbill, evening grosbeak, olive-sided flycatcher, winter wren, and kinglets, while the loss of older mixed-wood forests has had a similar impact for Swainson's thrush. The tree cavities of older or dying trees are also important wintering sites for small mammals, such as the northern flying squirrel.

The health of forests also depends on wildlife species. Plant pollination by hummingbirds, bats, and bees is critical to large numbers of flowering plants, including some willows and maples. The forest floor of the mixed forest is brightened by a host of beautiful wildflowers, including dogtooth violet, pink and yellow lady's slipper, and spring beauty, which are dependent on pollinators. And flowering plants in turn serve as food sources for some twenty-one species of butterflies in the eastern deciduous forests. Fruit-eating birds and mammals are also responsible for the dispersal of seeds and the perpetuation of some species of shrubs and trees, such as the flowering dogwood and oaks.

The diverse mixture of hardwoods and conifers supports around 222 species of breeding birds, making this region the second-richest within the whole temperate broadleaf and mixed forest biome. One hundred and forty species breed in the New England/Acadian Forest, of which the wood warblers, at twenty-three species, are the most common. Many, like the black-throated green warbler and Blackburnian warbler, show a preference for unmanaged, mature forests. The great attraction is insects: grubs, beetles, mayflies, moths, butterflies, blackflies, and mosquitoes begin to emerge in spring from the ground, from the bark of trees, and from meltwater, attracting a host of neotropical migrants, including redstarts, tanagers, and more than a dozen wood warblers. They join the populations of resident birds like chickadees, nuthatches, and siskins. Most of the forest birds are insectivores, whether they excavate bark, as woodpeckers do; glean insects from leaf and bark; or snatch them from the air, as the flycatchers do.

Not surprisingly, the forests that support the greatest diversity of birds are the ones that have the greatest structural diversity, thus providing many niches for insects. A large, multilayered forest with a rich groundcover layer, an understory, and a large canopy supports more birds than a small patch of forest with a single layer, such as a monoculture plantation.

This physiographic diversity not only attracts birds but also supports more

Bicknell's thrush

## THE NEWEST BIRD

ORNITHOLOGIST E.P. Bicknell first identified "a hitherto unknown thrush" on Slide Mountain in New York's Catskill Mountains in 1881. This songster was quickly consigned to taxonomic obscurity, however, when it was classified as a subspecies of the more widespread gray-cheeked thrush, which breeds in the boreal forests from Alaska to Newfoundland. More than a century later, DNA analysis demonstrated conclusively that it was a separate species, and in 1995 it became North America's newest bird—Bicknell's thrush. Its seeming obscurity relates in part to the remoteness of its preferred nesting sites, on offshore islands and on the virtual islands of the highest peaks of the Appalachians. It is the only native breeding bird from the northeastern United States and Maritime Canada and is documented as nesting in the Christmas Mountains of northern New Brunswick, and the Cape Breton Highlands, and on Mount Mansfield in Vermont.

Bicknell's thrush prefers dense nesting habitat in the subalpine spruce-fir zone just below the tree line, where high winds have stunted and shaped evergreens to form an almost impenetrable tangle—the krummholz effect. It is also found in the "fog forest" of stunted spruce along the coast, often with blackpoll warbler and fox sparrow. Because of its secretive character, fog-shrouded habitat, and well-concealed nests, it often has to be identified by its spiraling song, which one 19th-century observer praised as "a most ethereal sound, at the very top of the scale, but faint and sweet." Although it faces some threats to its "sky islands" in New England and in the adjacent Canadian provinces, its major challenge is on its Caribbean island wintering grounds, which include Hispaniola, Puerto Rico, and Jamaica. Only 13 percent of native forests still stand in the Dominican Republic and less than 5 percent in Haiti.

species of small mammals and amphibians. In New England there are approximately 338 common inland species. This diversity can be attributed to the wide array of habitats, which are a consequence of the elevation changes from the highest peaks of the Appalachians to the tidal marshes of the coastline, as well as to the latitudinal gradients, which bring northern and southern species together.

Characteristic large mammals of the region include black bear, red fox, snowshoe hare, porcupine, fisher, beaver, bobcat, marten, muskrat, and raccoon—a total of fifty-eight nonmarine species in the Atlantic Maritime ecozone of eastern Canada and northern Maine. A number of large mammals have been extirpated from the region, including elk, wolf, wolverine, and mountain lion, or eastern cougar, while the woodland caribou persists as an endangered population only in the Gaspé Peninsula, having disappeared from both New Brunswick and Nova Scotia by the late 1920s.

Hundreds of cougar sightings have been reported in the Maritimes since the 1980s, and since 2002 six hair samples have been collected from pheromone-equipped "hair poles" from New Brunswick and Quebec, and from a cougar hit by a truck in East Hereford in the Eastern Townships region of Estrie, Quebec, not far from the New Hampshire border. DNA analysis of the hair samples revealed that four of these animals, including the one struck in Estrie, were of South or Central American origin. These results suggest that these cougar were escaped or released animals or the offspring of once-captive animals. The remaining samples were of unknown ancestry or from

Flowering plants, such as the pink lady's slipper (right), are food sources for butterflies, such as the common wood nymph (left).

Fisher is one of the characteristic but rarely seen large mammals of the New England/Acadian Forest.

North American populations and might well be pet releases or the offspring of captive-bred individuals. Genetic analysis indicates that the North American cougar is a single race and that the eastern cougar, if it still exists, should be treated as the only valid subspecies occupying North America.

Regardless, wolves and cougars—both victims of hunting pressure—have been replaced ecologically by the eastern coyote as a top predator. Coyotes have made a century-long migration from western North America, arriving in northern New England in the 1930s and southern New England two decades later, and finally completing their cross-continental odyssey in the late 1970s by crossing the isthmus of the Tantramar Marsh into Nova Scotia. Since then, they have dispersed into all available habitats, even crossing the ice of the Strait of Belle Isle onto the island of Newfoundland.

The success of the coyote can be attributed to a number of factors. Coyotes are omnivores and have even been discovered to subsist on kelp and seaweed on offshore islands. Their preferred food is small mammals, especially red squirrels and snowshoe hares, but they also prey on mice and voles, and when necessary, like foxes, they eat fruit and insects. They are highly social animals and will hunt in packs or as mated pairs and will kill deer if snow conditions allow. Their two-parent system of raising young produces litters of five to nine. Coyotes' mythical cleverness may also contribute to their uncanny success. They work in pairs when stalking rodents, one approaching from a conspicuous position, the other stalking from behind. But extrinsic factors have also favored the spread of coyotes in the eastern woods. The extirpation, or suppression, of eastern Canadian wolves has allowed coyotes to move into territory where wolves once held sway. Although wolves and coyotes are archrivals, as canids they are able to interbreed, and it is believed that coyotes have picked up wolf genes on their travels. As a result, the eastern coyotes are significantly larger than their western counterparts.

*following spread*: Deer populations have expanded in response to a warming climate and increased browse in the wake of clear-cutting practices.

Deer have also moved to the north, where they replaced woodland caribou and have enjoyed uncommon success. Both moose and caribou were the most abundant large game animals in the eastern forest at the time of European colonization. Hunting seriously depleted both species, but caribou also suffered from the loss of their principal food, white reindeer moss, which was destroyed by widespread fires. Moose fared somewhat better until deer burgeoned in their traditional territory. The deer brought with them a parasitic brainworm, *Paralephostrongylus tenuis*, which, though harmless to them, was fatal to both moose and caribou. Deer populations exploded in the early 20th century with the large-scale removal of forests by clear-cutting, which released a surfeit of hardwood browse. Their numbers have grown to the degree that they are considered pests in some areas of the eastern forest, such as Pennsylvania, where they prevent forest regeneration by browsing seedlings. Black bears may also have been affected by changes in the forest composition. Bears would normally eat large amounts of fat-rich beech nuts before hibernation, but the loss of beech trees to disease may have affected the productivity of bears in northeastern forests.

### *Oak Domain: Northeastern Coastal Forest*

The northeastern coastal forest is native to the Piedmont Plateau and the Atlantic Coastal Plain, spanning seven states from northern Maryland to southern Maine. It is dominated by oak forests, including white, red, black, scarlet, and chestnut oaks. Northern red oak is also found scattered throughout the New England/Acadian Forest to the north and forms nearly pure stands on wetter sites, such as river floodplains; the other species are restricted to this region. The primary conifers in the region are white pine and hemlock, and pitch pine can be found on sandy outwash sites. White pine

were marked with the King's Broad Arrow as masts for British naval ships, but white oak was the species most sought after by local shipwrights when they came to the continent. Oaks of various kinds form a belt that spans the temperate regions of the globe. The great oceangoing ships of the Age of Discovery were framed entirely in oak and mostly planked in oak, and so it was in the New World, where oaks also were found. The USS *Constitution*, dubbed "Old Ironsides," had a hull of white oak so hard that it reputedly rejected British cannonballs during the War of 1812.

Oaks are rarely found in pure stands but are usually associated with hickories (shagbark, bitternut, mockernut, and pignut), and until the early part of the century they grew alongside the American chestnut. Chestnut trees grew straight and branch-free to a height of 16 meters (50 feet) or more, and averaged 2.5 meters (8 feet) in diameter. American chestnut was used for three centuries to build most barns and houses east of the Mississippi. The tree was such a vital part of this region, both commercially and ecologically, that the forests were referred to as oak-chestnut forests.

However, the accidental introduction in 1904 of the chestnut blight fungus into the Bronx Zoo from Chinese chestnut nursery stock decimated this once lordly species. The fungus was spread by the native chestnut bark beetle and expanded at a breakneck speed of up to 80 kilometers (50 miles) a year. By 1940, mature chestnut trees had been virtually wiped out, with an estimated 4 billion trees lost. In south and central New England, one out of four forest trees was a chestnut, so the loss has drastically affected the forest composition and ecology, though it is now difficult to determine exactly how. Fortunately, nuts that were produced before the adult trees died were able to germinate. Today, the species survives only as living stumps, or "stools," which produce shoots but seldom seeds before again being attacked by the fungus.

Oak acorns continue to be an important source of food for nearly one hundred species of birds and mammals in the northeastern coastal forest, including ruffed grouse, wild turkeys, woodpeckers, nuthatches, and ducks, as well as eastern chipmunks, white-footed mice, gray squirrels, white-tailed deer, and black bears. Densities of both white-footed mice and eastern chipmunks are directly correlated with densities of autumn acorns, a vital winter food for these small mammals. But white-tailed deer also consume large quantities of acorns to build up fat reserves for winter, depriving other wildlife of this food source.

Mature white oaks typically produce ten thousand acorns a year, but they produce bumper crops—as many as sixty thousand—every four to ten years.

Such a strategy seems intended to overwhelm the seed predators, improving the chances that some of the acorns will germinate and sprout, but the trigger for this strategy remains a mystery. Although animals can hinder oak regeneration when they consume too many acorns, they are also responsible for their dispersal.

Gray squirrels and eastern chipmunks bury acorns in leaf litter—a behavior called scatter hoarding—where many are likely never recovered and some germinate successfully. Blue jays also scatter hoard acorns, and it is believed that the blue jay was responsible for the rapid spread of oaks in eastern North America following the retreat of the glaciers at the end of the Pleistocene.

AMERICAN CHESTNUT

CATASTROPHIC DISTURBANCES TO New England forests from blowdowns and fire occur relatively infrequently, at intervals of about 1,150 and 800 years, respectively. Major storms such as the hurricane of 1938 and the ice storm of 1998 occur more frequently, on average every 150 years. Although they destroy mature trees, such storms do not appear to have much impact on the successional trends or overall species composition, since New England's pioneer species are resilient and survive in the understory.

Native Americans in southern New England used fire more frequently than those in northern New England to clear fields for planting, open the forest for easier travel, and drive game for hunting. Indians raised corn, beans, squash, pumpkins, and tobacco, moving their fields every eight to ten years when fertility declined. Beaver meadows, periodic fires on dry sites, and this pattern of shifting native agriculture maintained an open landscape along many of the major rivers as well as the coast. Burning usually occurred in the spring and fall and produced parklike woodlands with a rich herbaceous growth on the forest floor. In turn, this habitat encouraged an abundance of such game animals as deer, turkey, rabbit, and partridge.

By 1860, three-quarters of the arable land in southern and central New England had been cleared for pasture and farm crops. These lands began to be abandoned as early as 1825, however, when the Erie Canal opened access to the rich farmlands of Ohio and people began following the frontier west. In fact, nowhere else in the world was such a large area of land cleared and then suddenly deserted as in New England. Today, the situation is a complete reversal of what it was a century ago; 65 percent of southern New England and fully 90 percent of the northern region is now forested.

Consequently, it is the grassland and shrubland species of birds—bobolinks, grasshopper sparrows, and yellow-breasted chats—that have shown the most dramatic declines in recent decades. Bobolinks, for example,

require large expanses of grassland or old hayfields with a high level of litter cover and broad-leafed weeds in which to hide their nests. The loss of agricultural land to development and the change to earlier and more frequent mowing have contributed to the bobolink's decline in the Northeast. At the same time, resident forest birds such as the dramatic pileated woodpecker, which requires trees 50 centimeters (20 inches) in diameter in which to nest and roost, have significantly increased in numbers.

BOBOLINK

Southern coastal waders and shorebirds are expanding their range northward along the New England coastline—among them, great egret, snowy and cattle egrets, little blue heron, glossy ibis, American oystercatcher, and willet. Oystercatchers and willets may be recovering from intensive hunting, which was not curtailed until the 1920s, and the showy waders, likewise, may be rebounding from the era when they were hunted for their plumes, and their general population increase is now pushing them northward. Populations of southern forest birds—Carolina wren, red-bellied woodpecker, blue-gray gnatcatcher, worm-eating warbler, and orchard oriole, for example—have also burgeoned in southern Massachusetts and elsewhere in New England, again because abandoned agricultural land has been allowed to return to forest.

Even so, there are no old-growth forests in this region, with only small unconnected fragments of intact habitat remaining. The largest of these is the Cape May National Wildlife Refuge (3,220 hectares, or 7,956 acres), which protects a broad spectrum of coastal and inland habitats: salt marsh, forested uplands, cedar swamps, and maritime forest on adjacent barrier islands. This mosaic of habitats provides a critical refuge for a wide variety of resident and migrating birds—some 317 species of shorebirds, raptors, waders, and songbirds—as well as 42 mammals and 55 reptiles and amphibians. It survives as a bulwark against the sprawling development of malls and residences that is threatening to consume a quarter of this region's remaining wild habitat in the coming decades.

*facing page*: Southern coastal waders, like the little blue heron, are expanding their range northward along the New England coastline.

### *The Pine Barrens*

Within this larger northeastern coastal forest region, the Atlantic Coastal Pine Barrens is considered a disjunct region unto itself. Only 9,000 square kilometers (3,500 square miles) in extent, it includes the southern half of Long Island, the hook-shaped peninsula of Cape Cod, Massachusetts, and the famous New Jersey Pine Barrens. These sometimes scrubby forests of twisted and stunted trees usually arise on sandy soils and have depended on fire for their regeneration and maintenance.

"Sand-hills... are sometimes one hundred feet high and of every variety of form, like snow drifts, or Arab tents, and are continually shifting... Thus Cape Cod is anchored to the heavens, as it were, by a myriad of little cables of beach-grass, and, if they should fail, would become a total wreck, and erelong go to the bottom," wrote Henry David Thoreau in *Cape Cod*. Thoreau was substantially correct in crediting beach grass with anchoring the dunes of Cape Cod, but the pine-shrub oak forest also plays its part in preventing the cape's signature landforms from blowing away.

Scrub Oak is particularly abundant on the cape, where it clusters to form a dense and tangled understory. Most mature trees reach only 1 to 3 meters (3 to 9 feet), though some can achieve heights of 6 meters (20 feet). They often grow in association with other oaks, including post oak, which with pitch pine forms the wind-shaped, stunted forest typical of Cape Cod.

A katydid balances atop a ripening cranberry, one of the wild crops of the New Jersey Pine Barrens.

The largest and finest example of this pine-oak forest type in northeastern North America is found in the interior of New Jersey. The pine-oak forests and dwarf pine plains occur in an otherwise deciduous forest climate zone, primarily because the sandy soils have a low water- and nutrient-holding capacity that subjects the vegetation to periodic drought as well as to frequent fire. The Pine Barrens remains largely undeveloped despite its location in the otherwise densely populated urban corridor between Washington and Boston. On one side is Philadelphia and on the other lies New York; as *New Yorker* writer John McPhee has observed, "on a very clear night a bright light in the pines would be visible from the Empire State Building."

The region is referred to by the pejorative term "barrens," largely because the early settlers found the sandy soils unamenable to raising traditional crops. Even today, the two main agricultural commodities produced in the Pine Barrens are essentially wild ones: blueberries and cranberries.

Although the Pine Barrens remains sparsely populated to this day, vegetation has been shaped by human action, especially repeated fires, which in the historic period have occurred every ten to thirty years. Early explorers commented on the use of fire by the native Lenni Lenape, or Delaware, as a hunting technique. In 1632, off Cape May, the Dutch navigator David de Vries recorded that he

> ... smelt the land, which gave a sweet perfume as the wind came from the northwest, which blew off the land, and caused these sweet odors. This comes from the Indians setting fire at this time of the year, to the woods and thickets, in order to hunt; and the land is full of sweet-smelling herbs, as sassafras, which has a sweet smell. When the land blows out of the northwest, and the smoke is driven to the sea, it happens that the land is smelt before it is seen.

Two decades earlier, Henry Hudson also observed a "great fire" on the south Jersey coast, which he attributed to natives. The fires not only served to drive the deer, improve visibility, and aid travel by foot but also encouraged the growth of plants in the forest that attracted game.

Lumbering began in the early 18th century and first concentrated on Atlantic white cedar, which grew in dense stands in lowland "cedar swamps." Its light, straight-grained, and rot-resistant wood was deemed ideal for shingles. By the end of the 19th century, this resource had been exhausted by recurring fires and the cutting of immature stands. During the 1800s the pine forests furnished fuel wood for Philadelphia and New York, which had depleted their surrounding woodlands, and supplied coastal steam vessels with their source of power.

The greatest impact on the native forests was wrought by the production of charcoal, used in the nearby cities as domestic cooking fuel but more significantly as a substitute for coal in smelting iron, produced from limonite or bog iron precipitated from the tannin-rich, tea-colored local rivers. At the same time, a glass industry grew up that depended upon the high heats produced by burning pitch pine.

The history of the region, in short, is one of fire, both deliberate and accidental, but in either case the result of human activity.

The dominant tree in the Pine Barrens forest, the pitch pine, is well adapted to repeated fires. The thick bark protects and insulates the trunk from fire damage, and even if the fire reaches into the canopy and destroys the needles, the tree can recover, since its trunk and branches are supplied with dormant buds, which can quickly replace its seed-producing capacity. Trees that are repeatedly subjected to fire sometimes produce what are known as serotinous cones, which are tightly sealed by a resin, allowing the seeds inside to remain viable for years. If a catastrophic fire kills the tree, the heat will be great enough to open the cone and sow the seeds, which can take root and grow even in soil with no humus. Some pitch pines carry both serotinous and nonserotinous cones. The latter open mainly in winter and are scattered by the wind.

Fire also serves the pitch pine by killing competing hardwoods such as maples and oaks. Pitch pines and oak species usually occur together, however; pitch pines form the canopy on the drier sites, and oak is more dominant on the upland sites. The Pine Barrens ecosystem can be seen as a function of both moisture in the soil and soil texture, from fine to coarse, both of which influence the frequency and intensity of fires. Oak forests occupy the more nutrient-rich upland sites, which have finer soils that retain moisture. On coarser soils, which are drier because they allow rainwater to filter through them, fire is more frequent, and the pines and a different set of oaks become more dominant. Dwarf pitch pine and scrub oak form a pygmy forest under the driest, most fire prone conditions. And in the case of extreme disturbance, the pitch pine gives way to an oak-heath community.

PINE BARRENS TREE FROG

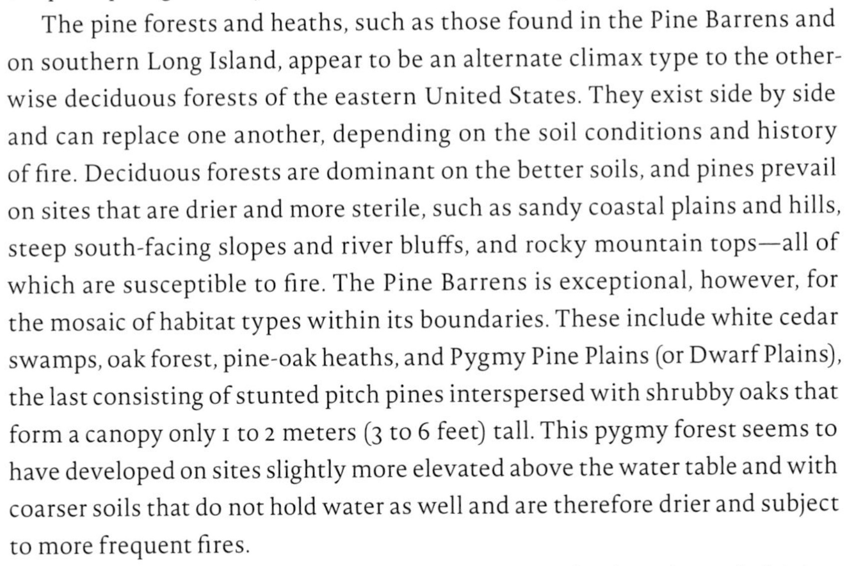

The pine forests and heaths, such as those found in the Pine Barrens and on southern Long Island, appear to be an alternate climax type to the otherwise deciduous forests of the eastern United States. They exist side by side and can replace one another, depending on the soil conditions and history of fire. Deciduous forests are dominant on the better soils, and pines prevail on sites that are drier and more sterile, such as sandy coastal plains and hills, steep south-facing slopes and river bluffs, and rocky mountain tops—all of which are susceptible to fire. The Pine Barrens is exceptional, however, for the mosaic of habitat types within its boundaries. These include white cedar swamps, oak forest, pine-oak heaths, and Pygmy Pine Plains (or Dwarf Plains), the last consisting of stunted pitch pines interspersed with shrubby oaks that form a canopy only 1 to 2 meters (3 to 6 feet) tall. This pygmy forest seems to have developed on sites slightly more elevated above the water table and with coarser soils that do not hold water as well and are therefore drier and subject to more frequent fires.

Although the Pine Barrens, in sharp contrast to the densely settled Atlantic Coastal Plain around it, has been drastically altered by human-induced fire, it remains largely undeveloped; it is altered but still essentially wild. The herpetofauna—amphibians and reptiles—is rich and varied for such a northern region. Fifty-eight species occur in southern New Jersey, including eleven salamanders, fourteen frogs and toads, eleven turtles, three lizards, and nineteen snakes. Latitudinally, it is at the northern limit of a number of southern species, among them the Pine Barrens tree frog and northern pine snake, which are both endangered in New Jersey. The most colorful might be the Pine Barrens tree frog, with its stripes of yellow and purple that run from the snout along the eye and side to the thigh, orange coloring on the inner surfaces of its legs, and maroon feet. It is particularly tolerant of the tannin-rich acidic waters that characterize the Pine Barrens waterways. On land, the northern pine snake finds the sandy soil ideal for excavating burrows, where it lays its clutch of two to twenty-four eggs in summer and hibernates during winter.

Breeding birds prefer particular habitats within the Pine Barrens, although many species are widespread in the upland habitats. The rufus-sided towhee is the most abundant species in the uplands, where its upbeat call—"drink your tea"—is often heard as it searches for food in the leaf litter. Other common upland birds include blue jays, Carolina chickadees, and black-and-white warblers. In wetter areas, the catbird and other scrub species take over, including yellowthroats and American redstarts.

NORTHERN PINE SNAKE

Generally, the harsh environment of the Pine Barrens, with its low habitat diversity, is not particularly attractive to mammals, but small mammals such as white-footed mice, woodland voles, red-backed voles, and meadow jumping mice are common and provide food sources for the corn snake, pine snake, and timber rattlesnake. White-tailed deer are plentiful throughout the Pine Barrens—so plentiful they often outstrip the carrying capacity, leading to winter starvation. River otter and muskrat also find a home in its many slow-moving streams and wetlands.

*Middle Atlantic Coastal Forest*

The Middle Atlantic coastal forest occupies the flat coastal plain from the eastern shore of Maryland and Delaware to just south of the Georgia–South Carolina border, giving way to the southeastern mixed forest, where the Atlantic Coastal Plain meets the edge of the Piedmont. The slow-moving blackwater rivers that snake across the flat terrain are famed for the towering bald cypress and gum trees that form river swamp or bottomland forests. Generally, the Middle Atlantic coastal forests are characterized by the most diverse assemblage of wetland communities in North America, including freshwater marshes, shrub bogs, Atlantic white cedar swamps, and tidally influenced bay heads.

These wetlands are critical habitats for a variety of reptiles and amphibians that require moist conditions to complete stages of their life cycles. The forests are also key to maintaining viable populations of thirty-seven species of southeastern songbirds as well as migratory waterfowl.

Tidal freshwater swamps, such as occur along the Pamunkey River in Virginia, furnish spawning and nursery areas for anadromous fish species and feeding and nesting areas for resident herons and migratory waterfowl. They are also home to a large number of herbaceous plant species, some of which are rare. The Pamunkey originates in the Piedmont, carrying red sediments from the uplands to the York River, one of four major rivers draining into Chesapeake Bay. At the time of European contact, the Pamunkeys, a tribe of the Powhatan Confederacy, had tribal villages on the bluffs along the river and used the tidal swamps as sources of game and fish, wild rice, and medicinal plants, which they continue to do.

*facing page*:
MAP: This range map of the Atlantic cedar (in dark red) marks it as a truly coastal species.

Flood-tolerant tree species—green ash, black tupelo (or gum), red maple, and American hornbeam—account for 96 percent of the canopy of the tidal swamps on the freshwater portions of the river. Half of the herbaceous productivity is attributed to green arrow arum and Asian spiderwort.

In frequently flooded tidal areas, an ash/tupelo community predominates, and bald cypress is sometimes an important component. In less frequently flooded tidal swamps, red maple and sweetgum dominate, defining a maple/sweetgum community. Diversity of both trees and understory species is higher on those sites subject to more frequent flooding.

Generally, river swamp forests, or bottomland forests, are dominated by bald cypress and swamp tupelo, but Atlantic white cedar swamps occur along blackwater rivers, which have organic soils underlain by sand.

Atlantic white cedar is a truly coastal species, restricted to growing within 250 kilometers (150 miles) of the Atlantic Ocean and the Gulf of Mexico. It occurs from southeastern Maine southward along the coastline to northeastern Florida and from the Florida panhandle along the Gulf Coast westward to Mississippi. Twenty-seven percent of Atlantic white cedar occurs in the North Atlantic states; New York State has the largest area of this species. It occurs in nearly pure stands, constituting over 75 percent of the stock, on poorly drained sites where the acidic soil is wet much of the time. In the northern Atlantic states—Massachusetts, Connecticut, New York, New Jersey, Delaware, and Maryland—it is often found with pitch pine and red maple. In the southern Atlantic states—North Carolina, South Carolina, and Virginia—black tupelo, white cedar, and red maple, as well as red bay, sweetbay, water tupelo, and pond pine make up sizable portions of the stands.

Historically, the largest concentrations of Atlantic white cedar occurred in North Carolina, but in the last two centuries intensive exploitation has reduced stands by 90 percent. Strong, lightweight, easily worked, and decay-resistant, cedar has many virtues that have made it highly sought after, and

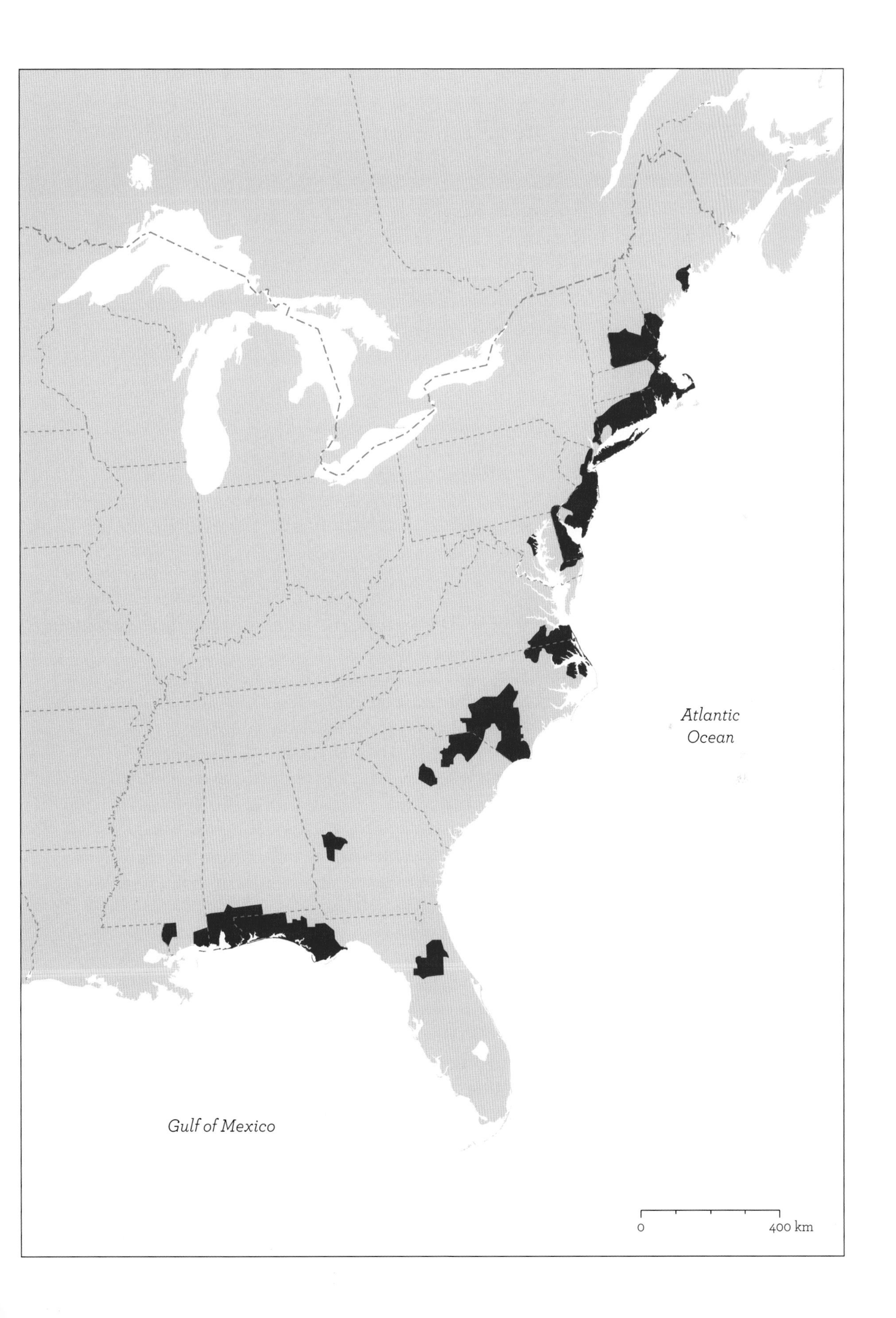

Atlantic
Ocean
Gulf of Mexico
0
400 km

today it is still preferred for boat planks, house siding, shingles, and fencing. Natural regeneration has often failed due to alteration of drainage patterns by the clearing of land for agriculture, the elimination of wildfires, and replacement by competing vegetation.

The Dismal Swamp was a name given in colonial days to Atlantic white cedar habitat that included 404,000 undrained hectares (998,000 acres) between the James River in southeastern Virginia and the Albemarle Sound in North Carolina. Today the Great Dismal Swamp National Wildlife Refuge, established in 1973, occupies a 43,000-hectare (106,000-acre) remnant of the former swamp. Cedar stands in the refuge support some of the greatest bird densities found in coniferous forests in the eastern United States, holding nearly twice as many birds as an equal area in the surrounding red maple–black tupelo forest.

ATLANTIC WHITE CEDAR

Parulid warblers are the dominant bird life in the Great Dismal cedar stands: prairie, prothonotory, hooded and worm-eating warblers, ovenbirds, and yellowthroats account for three-fourths of the breeding birds. Prairie and worm-eating warblers appear to be particularly dependent on the Great Dismal cedars. Cedar also seems to be a particularly important food source and habitat for wintering birds. On one occasion, ten thousand pine siskin were observed feeding in a single Great Dismal stand—the largest such gathering ever reported.

In the northeast, Atlantic white cedar is the preferred browse of white-tailed deer. Cottontail rabbit and meadow mouse also feed on cedar seedlings.

*Green Monster: The Boreal Forest*

The boreal forest has been compared to a green shawl draped over the shoulders of the continent. In fact, this great northern coniferous biome mantles the globe. Circumpolar in extent, it is constrained in the north by the treeless tundra and in the south by the temperate deciduous forest. It is vast, covering roughly 28 percent of the North American continent, stretching from Alaska to Labrador, where it crosses the water onto the island of Newfoundland. In Russia it is called the taiga, and its vastness was perhaps best captured by the great Russian playwright Anton Chekhov when he made an expedition across Siberia on horseback: "The strength and charm of Siberia does not lie in its giant trees and its silence, like that of a tomb, but in that only migratory birds know where it finishes. On the first day one does not take any notice of the taiga; on the second and third day one begins to wonder, but on the fourth and fifth day one experiences a mood as if one would never get out of this green monster."

## ATLANTIC COASTAL PLAIN FLORA

A GROUP of nearly fifty plants known as the Atlantic coastal plain flora are found in southwestern Nova Scotia, where they were first discovered in the 1920s by preeminent Harvard botanist M.L. Fernald. Most of these plants are characteristic of various wetland and heathland types of the eastern United States, where the ranges of some of these species extend northward only to New Jersey or to Cape Cod. Until recently, it was thought that these plants persisted during the last glaciation in a refugium off the Atlantic coast, when the sea level was 100 meters (300 feet) lower than today and the coastal plain was emergent. According to this scenario, as the glacier retreated the plants extended their range to Nova Scotia, where they became isolated.

It is now evident, however, that the Wisconsinan glacier advanced to the outer edge of the continental shelf of Atlantic Canada and that the emergent areas of Georges Bank and Browns Bank were separated from the Nova Scotia mainland by 150 kilometers (90 miles)—too far for the plants to have crossed a supposed postglacial "land bridge."

Instead, researchers now believe that coastal plain species reached Nova Scotia by stepwise migration through Maine, New Brunswick, and north-central Nova Scotia during a period of warmer climate. Subsequent cooling and vegetation change led to the loss of intermediate populations. Alternatively, coastal plain species may have reached southwestern Nova Scotia by long-distance dispersal from populations in the eastern United States, either by birds or by strong winds generated by tropical storms and hurricanes that track northward up the east coast of the United States, then often veer eastward toward Atlantic Canada.

Once there, the coastal plain plants found the conditions in southwestern Nova Scotia conducive to their permanent establishment. It is the warmest area of the province, with the longest frost-free period, and it has an exceptionally high density of lakes. Many of these plants are adapted to limited, nutrient-poor sites along lake and river shores, as well as bogs, fens, and estuaries. Often they are subject to the physical stress of wind, waves, ice-scouring, and fluctuating water levels. Such limiting factors work in their favor, however, because they do not compete well with more aggressive plants on more fertile and less stressful sites.

Eleven of these plants are protected by federal or provincial legislation as species at risk of extinction, and five of Nova Scotia's ACPF are globally rare: pink coreopsis, New Jersey rush, Plymouth gentian, Long's bulrush, and goldenrod. Human activities probably now pose the greatest threat to its survival. These include use of all-terrain vehicles, alteration of the shorefront habitat, eutrophication, and hydroelectric development, which can change the natural fluctuations in water level.

Despite the vast extent of the boreal forest, its flora is marked by simplicity. The overall impression of a boreal forest—say, looking down on it from a plane—is one of conformity. The few hardy trees of the "north woods" are the ones adapted to survive the harshness of a boreal winter. In a square mile of forest there might only be one or two species of trees, compared with hundreds in the same area of tropical rain forest. But those trees, unlike their tropical counterparts, can occupy a wide range of habitats—they are not adapted to fit a particular environmental niche.

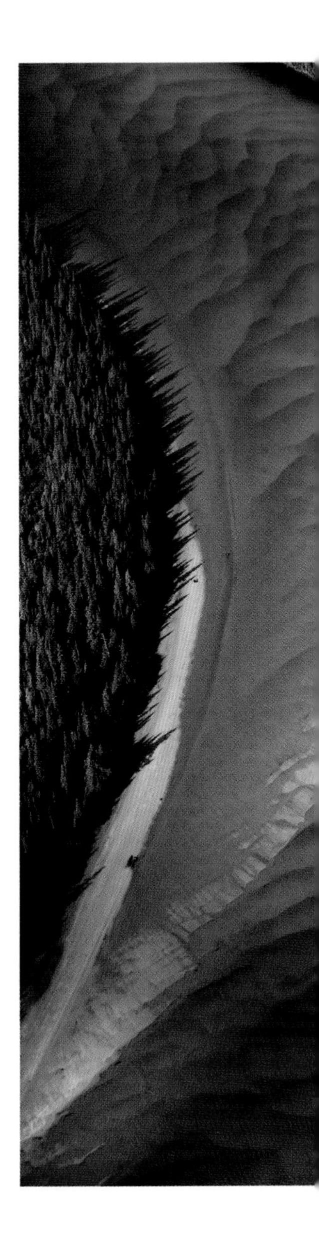

Black spruce emerge from a gray ground cover of caribou lichen along a great northern river.

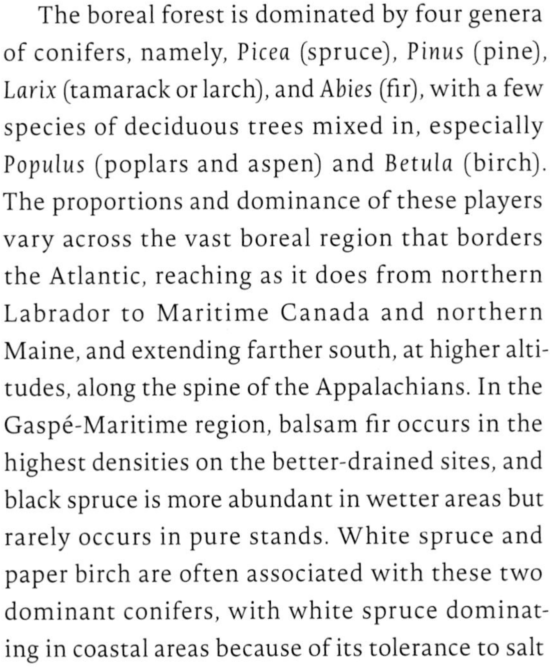

The boreal forest is dominated by four genera of conifers, namely, *Picea* (spruce), *Pinus* (pine), *Larix* (tamarack or larch), and *Abies* (fir), with a few species of deciduous trees mixed in, especially *Populus* (poplars and aspen) and *Betula* (birch). The proportions and dominance of these players vary across the vast boreal region that borders the Atlantic, reaching as it does from northern Labrador to Maritime Canada and northern Maine, and extending farther south, at higher altitudes, along the spine of the Appalachians. In the Gaspé-Maritime region, balsam fir occurs in the highest densities on the better-drained sites, and black spruce is more abundant in wetter areas but rarely occurs in pure stands. White spruce and paper birch are often associated with these two dominant conifers, with white spruce dominating in coastal areas because of its tolerance to salt spray. A moss-heath vegetation or barrens is typical of coastal areas exposed to high winds, such as Cape St. Mary's in Newfoundland, where the trunks of century-old shrubs are smaller than your little finger and will grow only to ankle height.

At other coastal sites in Newfoundland, wind-contorted balsam fir or white spruce grows so thickly it is impossible to penetrate it. It is sometimes possible, however, to walk on top of this hedgelike growth, which is the result of the krummholz effect, a term derived from the German for "crooked wood."

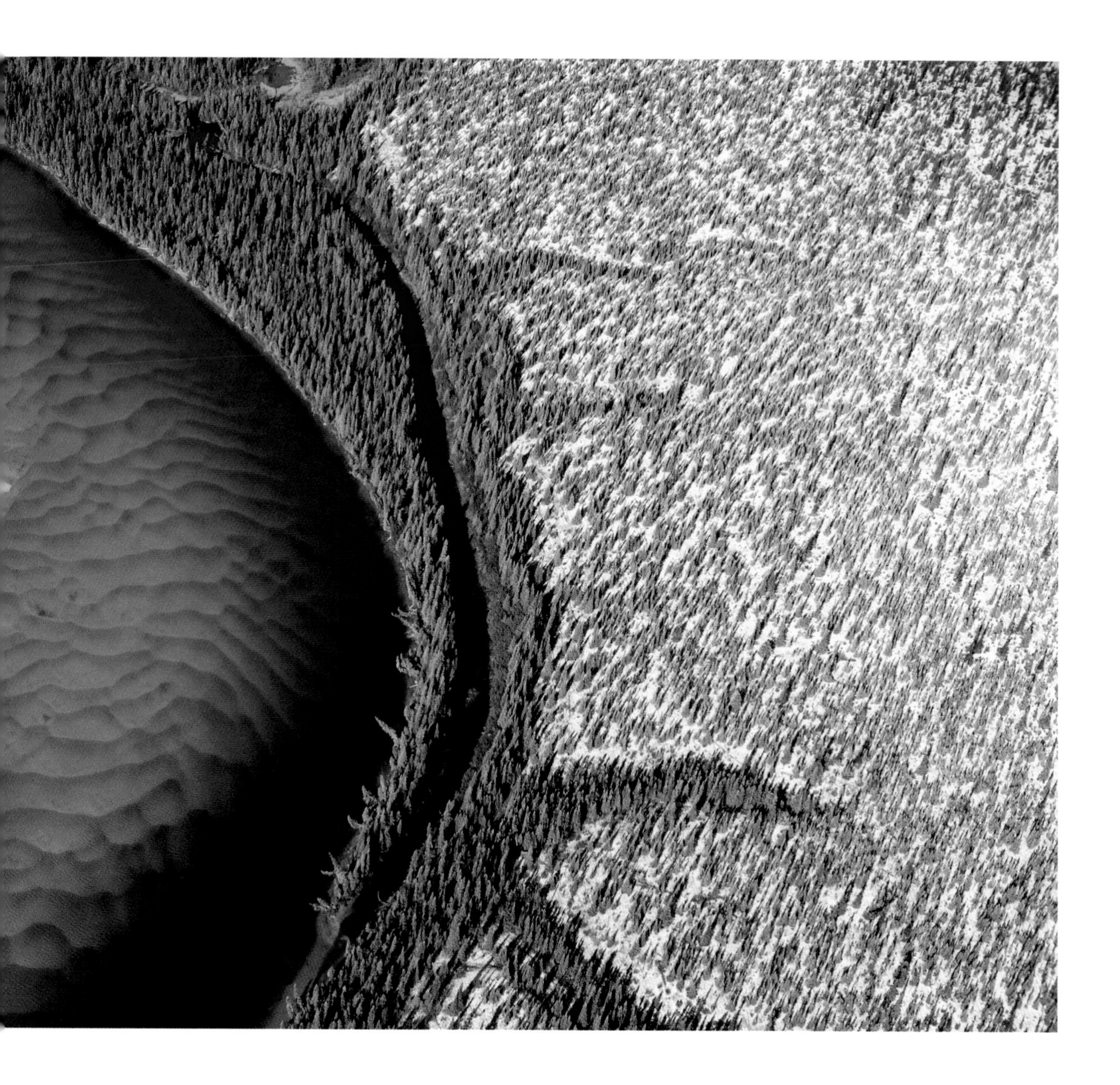

Krummholz is also common on the highest peaks of the Appalachians, such as Mount Carleton in New Brunswick and Mount Katahdin in Maine. These trees are usually less than 3 meters (9 feet) high and sometimes less than one meter (3 feet), but their bottom whorl of branches may extend out from the central stem more than 2 meters (6 feet). Dwarf shrubs and abundant lichen cover open ground on these alpine summits.

As you move farther north, black spruce replaces balsam fir as the dominant species until it takes over almost entirely on the Labrador-Ungava

Bearberry and Labrador tea brighten the groundcover of reindeer moss.

Peninsula. A.P. Low traversed this forbidding landscape in 1896 for the Geological Survey of Canada, and his acute observations still stand as the best description of this northernmost section of the Atlantic boreal forest:

> Black spruce is the most abundant tree of Labrador and probably constitutes over ninety per cent of the forest. It grows freely on the sandy soils . . . and thrives as well on the dry hills as in the wet swampy country between the ridges. On the southern watershed the growth is very thick everywhere, so much so that the trees rarely reach a large size. To the northward, about the edges of the semi-barrens, the growth on the uplands is less rank, the trees being in open glades, where they spread out their branches resembling white spruce. The northern limit of the black spruce is that of the forest belt; it and larch being the last trees met before entering the barrens.

Low also describes the understory of this northern forest, which no doubt provided him with much grief as he beat his way across the land:

> Throughout the forest belt, the lowlands fringing the streams and lakes are covered with thickets of willows and alders . . . Beyond the limits of the true forest, similar thickets of Arctic willows and birches are found on the low

grounds, but on the more elevated lands they only grow a few inches above the surface. In the southern region, the undergrowth in the wooded areas is chiefly Labrador tea (*Ledum latifolium*) and "laurel" (*Kalmia glauca*), which grow in tangled mats, from two to four feet high, and are very difficult to travel through . . . In the southern regions the ground is usually covered to a considerable depth with sphagnum, which northward of 51 degrees is gradually replaced by the white lichens or reindeer mosses (*Cladonia*).

By the time you reach the Torngat Mountains of northern Labrador, tundra consisting of lichens, mosses, and sedges cover half of the upland surfaces, whereas Arctic black spruce and a mixture of evergreen and deciduous shrubs survive on the more sheltered, wetter sites.

CONIFERS HAVE DEVELOPED a number of strategies to combat the cold and darkness of a northern winter. Year to year, they retain their evergreen needles, which carry out photosynthesis, and whenever the weather is warm enough, they can manufacture glucose. This strategy has been described by conservation ecologist J. David Henry in *Canada's Boreal Forest* as a "waste not, want not strategy," as opposed to the "easy come, easy go" strategy of the leaf-shedding hardwoods. Conifers also seem to have adapted to thrive on the thin, immature soils of the North, which have formed only in the last six to nine thousand years since the glaciers retreated. They are able to extract the nutrients in this organic-poor and acidic soil through a mutually beneficial relationship with mycorrhizae, fungi that are intimately interwoven with their root tissue. (Most plants of the taiga, including mosses, ferns, and angiosperms, take advantage of this ancient relationship.) The fungi supply the roots with nutrients and water, while the higher plant feeds the fungus with carbohydrates obtained through photosynthesis.

The most critical adaptation is to cold in the north, where temperatures can plunge below -40°C (-40°F). Hardy trees of the eastern deciduous forest, such as maples and oaks, can supercool the liquid within their cells even when temperatures drop below the freezing point, without the formation of ice crystals, which would damage the cell's proteins and organelles, killing the trees. Once the temperature drops below -40°, however, the liquid inside the cells of the hardwoods freezes. In the hardier conifers, these liquids are squeezed out through the cell membranes into the empty spaces between the cells, where the ice crystals that inevitably form at these temperatures can do no damage. A few hardwood trees typical of the boreal forest—paper birch, trembling aspen, and balsam poplar—also have this extracellular freezing capacity.

Much of the boreal forest in eastern North America arises on the granitic rocks of the Precambrian Shield, the core of the continent. Four types of vegetational communities grade into each other along a south-to-north latitudinal cline. In sequence, they are closed coniferous forest, open coniferous forest, forest-tundra, and treeless tundra. The trees gradually thin out until they disappear at the continental Arctic tree line and are replaced by tundra. This tree line is anything but straight, however, swinging north and south, with fingers of forest occurring along sheltered river valleys and giving way to treeless tundra on exposed uplands. In between are the ubiquitous lichen woodlands, a forest of widely spaced spruces, which appear as dark, steeplelike silhouettes against a pale ground carpet of lichens, interspersed with scattered northern shrubs and herbaceous plants.

The lichens of greatest importance in boreal forests, in both extent and sheer biomass, are the reindeer lichens. They grow on soils, both organic and inorganic, rather than on bare rock, bark, or decayed wood. One of their preferred habitats is rocky outcrops covered with the thinnest veneer of soil, where they have a competitive advantage over mosses and vascular plants. In general, lichens thrive in drier places, whereas mosses like wetter environments. But the greatest limiting factor for the growth of ground lichens is light. Few lichens can grow in the shaded habitat occupied by common boreal mosses, such as sphagnum moss. Lichen prefer thinly wooded habitat, where light penetrates to the forest floor. For this reason, lichens are pioneer plants and may be the first plants to predominate on a disturbed site—after a fire, for instance. In general, mosses predominate in moist oceanic climates, and lichens in drier continental ones.

### *Animals of the North Woods*

As in all ecosystems, the flow of energy through the boreal forest begins with plants and photosynthesis. But in Arctic and subarctic ecosystems, such as prevail in the more northerly parts of the boreal forest, short periods of intense primary production are followed by long, cold, dark periods when the accumulated energy cannot be passed through the food web.

Even so, northern forests support surprisingly large numbers of boreal animals. The mammals found here tend to be habitat generalists, such as wolf, red fox, beaver, muskrat, and deer mouse, which can adapt to changing conditions and thus have some of the widest distributions among North American mammals. Other typical boreal mammals include the red squirrel, nocturnal

A mink is on a slippery slope, serving as both predator and prey in the boreal ecosystem.

flying squirrel, woodchuck, porcupine, snowshoe hare, moose, black bear, and caribou, as well as smaller rodents such as mice and voles—which are largely herbivores. Among the major carnivores are coyote, lynx, marten, fisher, weasel, mink, and otter. Each of these species has its own food needs and preferences, which allows all of them to survive the whole year, even the winter, with its extreme cold and snow.

In summer, moose—one of the iconic animals of the northern forest—spend up to fourteen hours a day grazing on aquatic plants and browsing hardwood species such as willow, poplar, alder, and birch, whereas in winter they have been observed in Quebec to switch from birch to mountain maple. As omnivores, black bears vary their diet by consuming 70 percent green vegetation—leaves, berries, nuts, and seeds—and making up the balance with carrion and fish. Smaller mammals such as mink and weasel are important species in the northern food web because they are both predator and prey. Weasels depend on mice and voles for half of their food, and even more than that in winter. Mink prey on a wide variety of animals, including muskrat and hares. In turn, hawks, owls, foxes, and likely wolves and lynx, prey on weasels and mink.

One of the most fascinating aspects of the boreal forest food web is the population boom-and-bust cycle that affects many snow-adapted or snow-dependent species, including snowshoe hare, spruce grouse, voles, and predatory species such as lynx, fox, mink, and marten. Many hypotheses have been put forward to explain these persistent cycles, which have been repeating themselves at least as long as there have been observers of boreal ecosystems. An array of factors have been earmarked as contributing causes: sun spot cycles; weather, particularly as it applies to snow conditions, fire, and food availability; and predictably, disease and parasites. Some studies show that predators cannot remove prey fast enough to affect their population peaks, whereas others indicate that the number of prey controls the predator population—either way, a feast-or-famine situation. Another theory posits that an internal physiological mechanism, not an extrinsic factor, kicks in to restore balance when a population gets too high or too low.

Perhaps the most fascinating and studied of these cycles is the one linking lynx and snowshoe hare numbers. This predator-prey relationship was first documented in the 1930s, when numbers of pelts of snowshoe hare and lynx received by the Hudson's Bay Company were analyzed. It was clear that hare populations built up to a peak and then crashed, on average every ten years. The cycle of the hare's principal predator, the lynx, lagged a year behind the hare's cycle, which also affected other species such as red fox and great horned owl. Because fluctuations in its numbers affect the populations of several other species in the ecosystem, the snowshoe hare has been called a keystone species. This cycle is repeated in synchrony across the continental swath of the boreal domain, although it lags by a year or two in Newfoundland, where the hare was introduced in the late 19th century. Even here, however, the cycle began repeating itself shortly after the hare's introduction, as it had, and does, elsewhere in the boreal world.

*facing page*: The lynx is a top predator in the boreal forest.

The persistence of this cycle is believed to be related to the simplicity of the northern ecosystems. In the case of the hare and lynx, it is close to being a one-prey, one-predator system. Because there are few if any other prey to support the lynx, once the hare population goes into decline, the lynx population must soon follow. Since it takes three to four years for hare numbers to recover—approximately one quarter the length of the whole cycle—the predator and prey populations do not reach equilibrium and the cycle is maintained. But the primary causes of the crash and the lag time have been subject to a number of theories and are still open to debate.

The interaction of food supply and predation appears to cause the die-off of hare populations, which is worsened by weather factors. During the peak population, predation exacts a heavy toll—83 percent of mortality, in one study—even when food supplies are adequate, driving the hares into decline. Vegetation seems to play a role as well. At peak population levels food resources are somewhat reduced, forcing hares to feed in areas with less shrub cover and therefore making them more vulnerable to predation. At the same time, snowshoe hares become stressed by the failed attempts at predation by coyotes and lynx, which are successful only 30 to 40 percent of the time. Increases in the level of stress hormones in the hares' blood stream are thought to permanently alter their reproductive physiology, with the consequence that litter size declines. These hormones are then passed on to the next generation through the placenta of the stressed females. The lag of three to four years in the cycle may be related to the fact that it takes that long for enough females without these stress-altered physiologies to begin reproducing at normal levels.

ANYONE WHO HAS spent time in the northern woods is only too aware of the torment of biting insects experienced there. But this miasma of mosquitoes, blackflies, and other insect life, which drives warm-blooded creatures to distraction, also attracts millions of feathered migrants, many of which winter in the neotropics of the Americas. These insects provide the protein necessary for reproductive success. Among the long-distance migrants are a broad range of bird species, including osprey, peregrine falcon, common nighthawk, alder and least flycatchers, Swainson's thrush, red-eyed vireos, and more than twenty warbler species, including Tennessee, yellow, chestnut-sided, Cape May, palm, and bay-breasted warblers. Density of breeding pairs is very high, ranging from three hundred to six hundred pairs per square kilometer in balsam fir, the richest conifer habitat in the boreal region. The obligate insect-eaters spend only three months in the north hawking their airborne prey, whereas insectivores that also eat seeds and fruit might linger for five months. Of the three hundred species that travel distances to reap the advantages of the boreal forest's smorgasbord, only one-tenth, or thirty, winter there. These hardy winter birds include the raven, black-capped chickadee, blue jay, spruce and ruffed grouse, woodpeckers, grosbeaks, and owls.

IN THE FAR north, changes to the forest composition and the animals living there are likely to come about principally through climate change. But farther south, as we have seen in the New England/Acadian Forest, human activities have reduced the types of trees and the overall age of the forest—and therefore the kinds of animals living there. For the near future, mammals best adapted to older forests are likely to decline, while species using younger forests, such as white-tailed deer, might well continue to increase as they have done during the last century. It has been shown that an increase in mean summer temperature of as little as one degree might destabilize predator-prey relationships, and as a result lynx populations could switch from cyclic to noncyclic behavior. Climate change may adversely affect mammals that need snow, such as the snowshoe hare and lynx, but positively affect those that have difficulty surviving long, hard winters, such as white-tailed deer and hibernating bats.

A snowshoe hare hightails it for cover. Its population is in lockstep with predators like the lynx.

Whereas southern species might move north, as the white-footed mouse already has done, expanding its New England range since the 1950s into northern coastal Maine, it is unlikely that boreal species, such as lynx and northern bog lemming, will move southward. Birds are more mobile than mammals, and the composition of bird life in the Maritime Provinces and northern New England has been in a state of constant flux since European colonization, with fully 22 percent of known breeding bird species having arrived after European settlement. Grassland species, such as bobolink, have significantly declined as farmland, which peaked in the late 1800s to early 1900s, has reverted to forest. At the same time, the loss of big trees has caused the decline of birds dependent upon them, such as the red crossbill and pine grosbeak. The only prediction possible is that future changes to the landscape will be accompanied by the historical pattern of near-constant change in bird life.

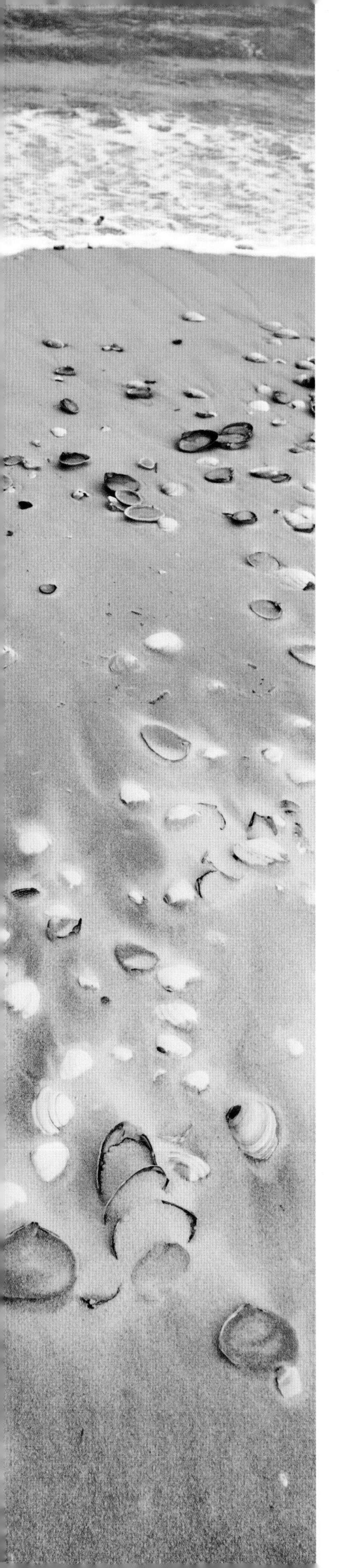

# 4

# BETWEEN THE CAPES

## *The Mid-Atlantic Bight*

LINES OF SHELLS—SURF clams, cockles, and bay scallops—shine under the strong, pewter-hued light streaming through a bank of dark cloud hanging ominously over New Jersey's Island Beach State Park. The carcass of a small shark is partially covered by sand at the wrack line. Just offshore, a common loon cruises parallel to the coast, its seemingly star-checkered body periodically dipping below the waves as it feeds; farther out, gannets are plunge-diving, sending up tiny explosions of spray as they enter the sea. Surf casters string out along the beach, trying their luck for striped bass and bluefish, while small flocks of great black-backed gulls huddle against the strong winds whipping the sand into the air and driving it, glittering, inland.

The Mid-Atlantic Bight is bounded by two capes—Cape Cod in the north and Cape Hatteras in the south. A distinctive feature of this ecoregion is the barrier islands, like Island Beach, New Jersey, ribbons of sand that unfurl for more than 3,000 kilometers (2,000 miles). This barrier island system actually begins north of Cape Cod, at the mouth of the Merrimack estuary—Plum Island and Salisbury Beach in Massachusetts—but it finds its most dramatic expression in Massachusetts in the dunes of Cape Cod, Martha's Vineyard, and Nantucket Island. The shorelines of Rhode Island and Long Island, New York, are also guarded by barrier beaches and spits consolidated from glacial sediments. Then, beginning with the unglaciated coast of New Jersey, an almost

< A treasure trove of shells collects in the swash zone of Island Beach, New Jersey.

The bright eye and bill of the oystercatcher probes for invertebrate prey in the shallows.

*facing page*: MAP: The Mid-Atlantic Bight.

continuous chain of barrier islands strings along the Atlantic coast and reaches around Florida's panhandle into the Gulf of Mexico.

Built up and shaped by the pounding surf, these islands not only protect the land behind them but furnish a mosaic of wildlife habitats, from energetic ocean beaches to scrubby maritime forests in the interior of the islands, to more tranquil but teeming salt marshes and lagoons on their landward side.

Rising sea level since the last glaciation has created not only barrier islands but also a series of estuaries behind them, where freshwater drains into the embayments formed between the island and mainland. Long Beach and Barnegat Bay in New Jersey, the Outer Banks of North Carolina, and Albermarle and Pamlico Sounds are prime examples of these paired barrier island–estuarine systems along the Mid-Atlantic coast. The other major estuaries, Delaware and Chesapeake Bays, were created as the rising sea flooded river valleys that had been gouged out by raging rivers from glacial meltwater during the last ice age, when sea level was as much as 100 meters (300 feet) lower.

The more southerly Chesapeake Bay is in a class by itself as the largest estuary in the United States. Half of its water volume derives from the Atlantic Ocean, matched by fresh waters that flow toward the sea from a massive drainage basin of 166,000 square kilometers (64,000 square miles), including parts of New York, Pennsylvania, West Virginia, Delaware, Maryland, and Virginia. The southern New England portion of the coastline includes Long Island Sound and Narragansett Bay.

The Gulf Stream begins to veer away from the coast north of Cape Hatteras, but occasionally a northward-bending meander breaks off the stream to form an eddy, or ring, that traps warm Sargasso Sea water in its center. These warm core rings are typically 100 to 200 kilometers (60 to 120 miles) across and can extend to a depth of 1,500 meters (4,900 feet). They are the only mechanism whereby warm tropical water can be transported from south of the Gulf Stream to north of it. Although these warm waters from the Sargasso Sea have relatively low levels of biological production, they sometimes transport tropical fish and sea turtles into the cool waters of the continental shelf in the Mid-Atlantic Bight.

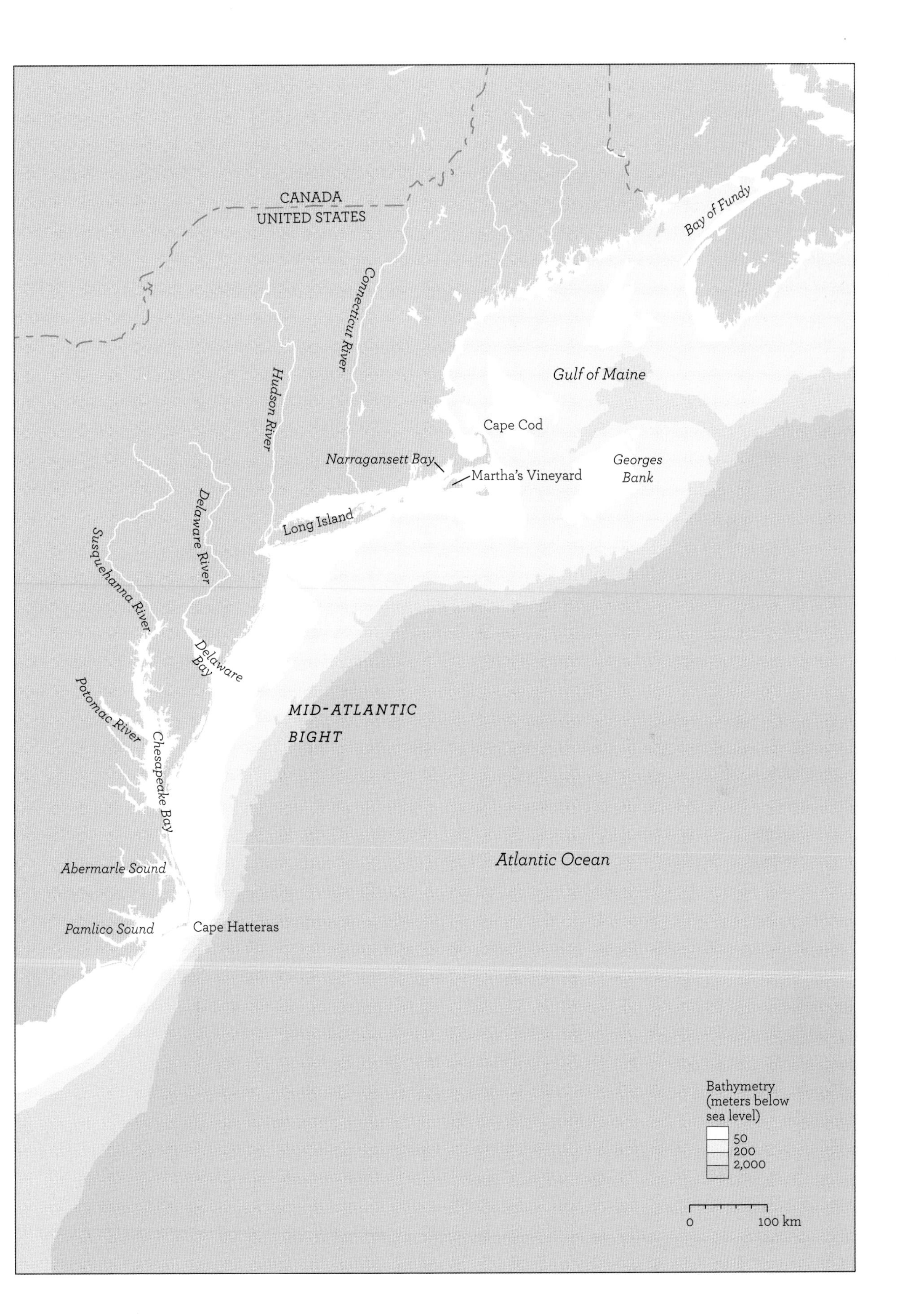

CANADA
UNITED STATES
Bay of Fundy
Connecticut River
Hudson River
Gulf of Maine
Cape Cod
Narragansett Bay
Martha's Vineyard
Georges Bank
Long Island
Delaware River
Susquehanna River
Delaware Bay
Potomac River
Chesapeake Bay
MID-ATLANTIC BIGHT
Atlantic Ocean
Abermarle Sound
Pamlico Sound
Cape Hatteras
Bathymetry (meters below sea level)
50
200
2,000
0
100 km

*Ribbons of Sand*

The barrier beach systems of the Mid-Atlantic, whether islands or spits attached to the mainland, have a common characteristic. They are adapted to the combination of physical forces that constantly challenge and seek to destroy them, including rising sea level, storms, wave-driven currents, and wind. Not only have the islands and their ecosystems withstood these powerful stresses without being destroyed, it is these forces that shape and maintain them over time.

Just as they were created by rising sea level, barrier islands must continue to adapt to the encroachment of the sea. Barrier islands are ever on the move toward the land and in the process undergo continuous change while, paradoxically, remaining the same. The position of the island migrates, but the ecological units of the island—dune, maritime forest, salt marsh, and tidal flat—are retained in their regular sequence, just displaced landward. The beach acts as a living thing, adapting to the environmental stresses imposed on it by rising sea level while at the same time retaining its essential character.

The movement and transformation of the island is often initiated by a major storm that causes waves to wash over the island or to push through low areas between the dunes, carrying with them large loads of sand, which are deposited when the storm surge loses its energy. Sand is deposited either on top of the salt marsh or, if the surge and tides are great enough, in the sound or lagoon behind the island itself. The deposits are called overwash fans to describe the shape of the sand deposit. Overwash is a regular event in the Outer Banks of North Carolina, which consists of low-lying islands that lack high, well-stabilized dunes. It is a less frequent, if not uncommon, event in areas such as Cape Cod, where dunes rise to dramatic heights.

When the overwash process delivers the sand into the bay behind the island, the landward movement of the island is clearly extended, although parts of the fan may remain underwater. Often, however, the overwash fan is deposited on top of a living salt marsh. This kick-starts an ecological response in which the barrier island essentially rolls over on itself, moving landward but maintaining the succession of habitats from its seaward to its landward edge. American beach grass colonizes the new sand deposits, usually beginning on the landward side of the overwash fan. Then prevailing onshore winds deliver new sand from the seaward side of the overwash deposits, which stimulates growth of the grasses and stabilization of the new dune. Eventually, a new, high foredune is created with extensive dune fields and other habitats, such as maritime forest and salt marsh, behind it—and the whole island is displaced toward the land. Dunes, and therefore islands, can also migrate when the wind moves sand and dunes toward the back of the island.

The landward displacement of the barrier island, whether it occurs through a dramatic incursion of the sea or by the gradual but steady force of the wind, continually recreates the same succession of habitats. From the beach (or swash zone) landward, in simplified terms, these habitats are dunes, shrub thicket and maritime forest, salt marshes, and tidal flats, with a sheltered lagoon or bay separating the island and the mainland. In each case a community of plants and animals has adapted to the demands and rewards of the habitat.

Beginning in the swash zone, where the turbulent surf crashes ashore, we are likely to encounter a number of well-adapted and visible creatures that live in the subtidal area bordering the beach and on or under the beach surface itself. The most impressive shells on the beach belong to the Atlantic surf clam, the largest clam living along the Atlantic coast, where it can be found from the Gulf of St. Lawrence to Cape Hatteras. The triangular shell can grow to a maximum of 23 centimeters (9 inches) during the mollusc's twenty-five- to thirty-year lifetime. Surf clams are prolific, maturing sexually as early as three months of age, and older individuals can broadcast as many as 13 million eggs at a time. They burrow into the sand at the lowest intertidal level, but most make their home below the tide line to depths of 60 meters (200 feet). Surf clams are filter feeders equipped with two siphons, one that draws water across a large set of gills, which extracts food and oxygen, and the other used to expel water and waste. These siphons are relatively short, however, so the clams can burrow only just below the surface. As a result, they are vulnerable to predators such as gulls, which pluck the clams exposed by the pounding surf and then break them open by dropping them on hard surfaces such as rocks, or even nearby parking lots. The clams have two large adductor muscles that hold the shells together, and the sweetness of these muscles makes

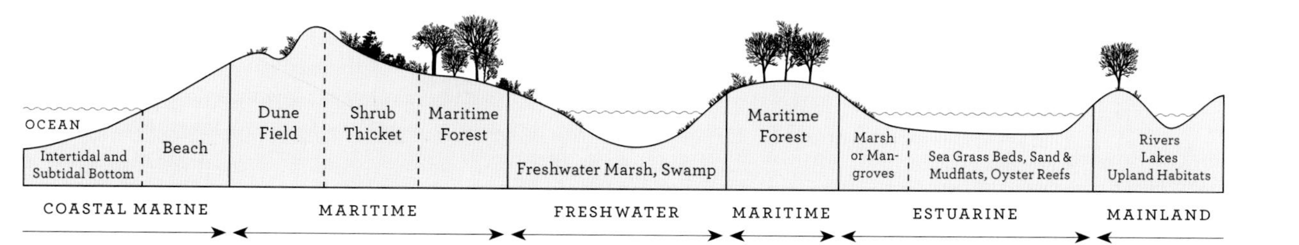


A cross-section of a typical barrier island and nearby mainland showing the diversity of habitats.

them the target of a clamming industry. Moon snails also seek out surf clams and open them expertly, first softening the calcium carbonate shell with an acidic secretion and then drilling a hole through the shell with their rasplike radula, which acts like a conveyor belt equipped with hardened teeth.

The surf clam is the largest clam living along the Atlantic coast, reaching a maximum length of 23 centimeters (9 inches).

Another invertebrate in the swash zone is the diminutive mole crab. These peanut-sized crustaceans are cryptically colored to blend in with their sandy environment, with only their eye stalks and two pairs of antennae poking up like periscopes above the beach. They emerge from their burrows as waves break over them, then deploy the feathery antennae to capture diatoms and other algae from the backwash of waves. As they are tumbled back toward the sea, they latch onto the bottom and quickly bury themselves using a powerful pair of paddles known as uropods, found on either side of the heavy triangular tail, or telson. The uropods scull like tiny propellers, allowing the tiny crab to seemingly disappear into the sand between waves before beginning its feeding cycle again. The antennae, which have as many as 150 joints, roll up and deliver the microscopic food items to the mouth of the crab, which swallows its food whole. It follows the moving tide up and down the beach but feeds only as the waves recede.

Mole crabs mate in spring and early summer, when the female carries bright orange-yellow egg masses cemented to her legs under the abdomen. Male mole crabs are only half the size of a full-grown female, which may grow to 2.5 centimeters (1 inch). During the mating cycle, two or three males cling to the female's back with specialized legs equipped with suction discs. Female mole crabs live little more than a year and the males less than that, either dying off after mating or, in some cases, undergoing a sex change into female crabs.

Mole crabs themselves are important prey, vulnerable to the probing bills and keen eyes of shorebirds skittering along the surf line, and to juvenile fishes such as Atlantic silverside, spot, and white mullet that are able to navigate the shallow waters of the swash zone. South of Cape Henlopen, Delaware, another crab emerges from its small burrows to prey on them as well. The well-named ghost crab, or sand crab, is primarily a nocturnal predator; it takes shelter in its tiny burrows from the sun and birds during the day but prowls the swash zone at night.

ALTHOUGH BEACHES ARE ever attractive to human beings as places of rest, recreation, and solace, they are extremely stressful environments to which few marine creatures are adapted. The constant pounding of waves—some eight thousand every day—means that plants and animals that need a solid surface to cling to cannot survive there. Those organisms that can adapt to the ever-changing environment of shifting sands often do so in high numbers. Many of these survivors are not visible to the casual beachcomber. This community of beach dwellers is known as meiofauna, and at 42 to 500 microns in size, they are small enough to live in the spaces between the sand grains, where they can attain incredible densities of 2 billion individuals per square meter. Permanent members of the meiofauna are the harpacticoid copepods and ostracods (seed shrimp), and segmented worm and crustacean larvae often spend part of their lives in this micro-world between the grains. The most dominant are the nematodes, or roundworms, usually less than 1 to 2 millimeters (1/16 of an inch) long and tapered at both ends. Most are thin enough to thread their way between the sand grains, where they graze upon microorganisms and sometimes other nematodes.

MOLE CRAB

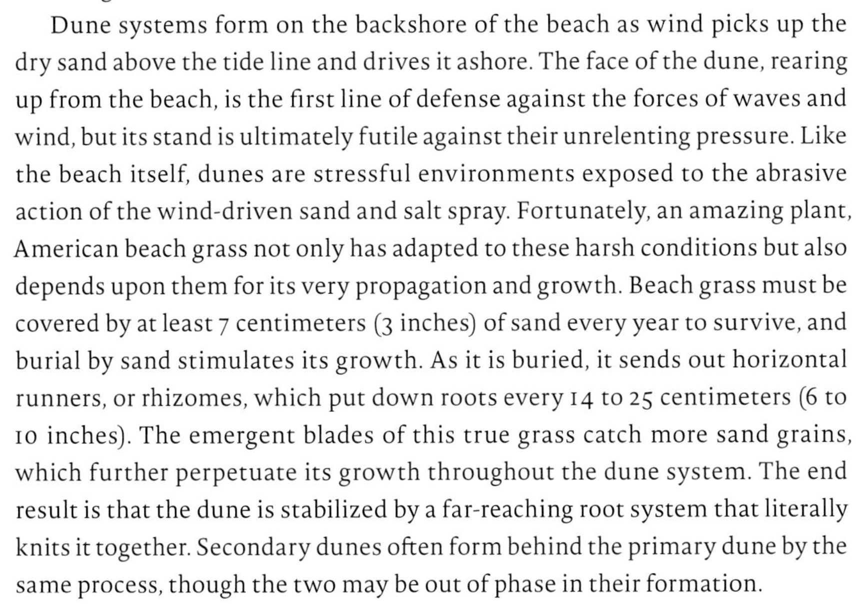

Dune systems form on the backshore of the beach as wind picks up the dry sand above the tide line and drives it ashore. The face of the dune, rearing up from the beach, is the first line of defense against the forces of waves and wind, but its stand is ultimately futile against their unrelenting pressure. Like the beach itself, dunes are stressful environments exposed to the abrasive action of the wind-driven sand and salt spray. Fortunately, an amazing plant, American beach grass not only has adapted to these harsh conditions but also depends upon them for its very propagation and growth. Beach grass must be covered by at least 7 centimeters (3 inches) of sand every year to survive, and burial by sand stimulates its growth. As it is buried, it sends out horizontal runners, or rhizomes, which put down roots every 14 to 25 centimeters (6 to 10 inches). The emergent blades of this true grass catch more sand grains, which further perpetuate its growth throughout the dune system. The end result is that the dune is stabilized by a far-reaching root system that literally knits it together. Secondary dunes often form behind the primary dune by the same process, though the two may be out of phase in their formation.

Other salt-tolerant plants, such as seaside goldenrod and sea rocket, may establish themselves on the foredune and the crest of the primary dune. The farther inland one progresses, the more terrestrial the vegetation becomes. On the leeside of the primary dune, which is more sheltered from salt spray, beach heather forms islands of yellow-blooming flowers in early summer—so-called heather holds—often under the canopy of beach grass. Shrubs such

Beach plum and heather "holds" brighten and anchor a Cape Cod dune.

as beach plum, highbush and lowbush blueberry, and black cherry may also take hold, though they are often pruned on the seaward side by wind and salt spray. Many of these salt-resistant plants, such as northern bayberry and wax-myrtle, have waxy leaves that protect the plant's tissues from the damaging effects of salt. These dense thickets of berry-producing shrubs and trees are rich sources of food for a variety of migrating songbirds, such as mockingbirds, brown thrashers, catbirds, and yellow-rumped, or myrtle, warblers. Virginia creeper winding through the dense growth of shrubs also helps to stabilize the dune systems with its root system, which can attain lengths of 16 meters (50 feet).

The thicket also provides protection for the growth of the larger trees that form the maritime forest, a dense, junglelike environment. This rare forest type is restricted to barrier islands and other exposed coastal sites. The trees that grow here cannot grow in saline waters but are tolerant of salt spray and include pitch pine, Atlantic white cedar, red maple, American holly, southern red oak, sweet bay magnolia, and willow oak. Freshwater wetlands develop in depressions among these patches of forest and provide habitat for muskrat, frogs, turtles, and aquatic insects. Dragonflies often congregate on the islands in massive swarms during their fall migrations, using the straight lines of the barrier islands as navigational markers. In addition, migratory songbirds depend on coastal maritime forests for food and cover during their fall migration, when huge flocks, or "tornados," of swallows descend on bayberry bushes, bending them to the ground with their weight.

The maritime forest is also where most terrestrial vertebrates are to be found. Toads, frogs, and snakes can easily reach the islands, as do skinks and salamanders, though with greater difficulty. One amphibian that commonly occurs in beach habitat from New England to the Gulf Coast is Fowler's toad, which is slightly smaller than the more familiar American toad found inland. The isolation of the islands favors genetic drift—as Darwin in the Galápagos Islands and Wallace in the Malay Archipelago first deduced. This phenomenon, which can result in the loss of a particular gene variant in small populations, has produced separate species or subspecies of amphibians, reptiles, and even mammals on some islands. The mammals, in fact, have adapted best to the island habitat and include herbivores such as squirrels, mice, rabbits, and deer, along with predators such as foxes, mink, and otters.

Between the island and the mainland shore are the salt marshes and back bays or lagoons, which are among the most productive habitats, acting as nurseries for local fish species and feeding and breeding grounds for birds. This is an especially important area for waterbirds and waterfowl which breed and/or winter here. Egrets, osprey, and herons have their nesting sites in the trees during the summer; in winter, a variety of waterfowl, including canvas backs, use the lagoons as food rich and sheltered areas. But the lagoons are especially important to overwintering brant.

The yellow-rumped, or myrtle, warbler (left) is attracted to the berries of the maritime forest, such as those produced by the namesake myrtle bush and holly (right).

*Sex and Gluttony in Delaware Bay*

Delaware Bay is host to two remarkable migrations that come together, precise as clockwork, along its beaches and marshes: that of primitive arthropods known by the colloquial name horseshoe crab, which come ashore from the deep waters of the bay and beyond to lay their eggs, and huge flocks of shorebirds, notably red knots, ruddy turnstones, sanderlings, and semipalmated sandpipers that greet the crabs after making a spectacular flight from Patagonia, at the southern tip of South America.

For many years I wanted to witness this spectacle, which is timed to the full or new moons in late May. To get there I first had to run the gauntlet of seemingly endless strip malls in northern New Jersey, a place where nature and any semblance of its life-giving processes have long ago disappeared under asphalt. Finally—and mercifully—the Garden State Parkway opens to surprisingly extensive green vistas of salt marshes on the ocean side and, inland, tall dark silhouettes of pitch pines, the signal species of New Jersey's famed Pine Barrens. I arrived in Cape May, the historic seaside resort town at the tip of the peninsula, on the night of the full moon. Before sunset, I drove out to some of the nearby beaches to find out whether the crabs had come ashore. Disappointingly, the beaches seemed deserted, and locals opined, "No, not many crabs yet."

A great egret feeds its hungry young.

## SHARING THE BEACH

BEACH-NESTING BIRDS along the Atlantic coast are threatened by increased commercial and residential development and recreational use of beaches. The most endangered is the piping plover, a sparrow-sized shorebird that nests on sandy beaches from Newfoundland to North Carolina. Today there are fewer than eighteen hundred along the entire coastline. Piping plovers commonly nest with least terns—the smallest of the terns, at a mere 23 centimeters (9 inches) in length—another beach nester under pressure. A third beach nester, the black skimmer, is threatened in some areas of its range. Each of these species faces similar pressures: human disturbance, harassment from pets, especially dogs, and beach developments that attract predators such as raccoons, skunks, and foxes. The species partition nesting sites on the outer beaches, each selecting slightly different areas. Oystercatchers, piping plovers, and least terns nest at the foot of the dunes, just above the tide line. They choose these open areas so that they can have an unobstructed view of any approaching predators. At the same time, they must choose a site above the high-tide wrack line to prevent their nest and eggs from being washed out to sea.

Least terns share their beach nesting sites with recreational users.

Larger beach nesters, like common terns and black skimmers, choose sites in the swales between the dunes, where the beach grass and other beach plants provide some shade and shelter from sun and rain as well as predators. They nest in larger numbers than the least terns and so provide their own defense of territory. On the highest ground, where shrubs and bushes overlook the dunes, herons, egrets, and ibises jockey for their nest sites. The largest species, such as the great egret, take the highest and safest sites, and the smaller snowy egrets, night herons, and glossy ibises settle for the more vulnerable lower sites.

Early the next morning, I made way for the once-sleepy fishing village of Reed's Beach, which has become a prime site for shorebird watchers in spring—and there, strung along the crescent beaches, beside the modest fishers' homes and cottages, were long lines of shorebirds, chirring with delight and seemingly oblivious to our presence as they went about the business of probing the sands for horseshoe crab eggs. I should not have been surprised that one day there were no birds or crabs in sight and the next they were there by the tens of thousands. Such is the nature of the biological clocks that dictate the movements and behaviors of many species, arguably including our own, for here I was timing my arrival from 1,500 kilometers (900 miles) away to arrive at a remote beach the morning after the full moon.

As a longtime resident of the Bay of Fundy, where more than a million shorebirds stop over to feed in late summer on their return journey from Arctic breeding grounds, I had seen impressive massings of shorebirds before. But I was surprised and delighted by the bold markings of the shorebirds assembled on the beach. I was used to seeing the shorebirds at a time of the year when their summer feathering was beginning to fade into the drab and muted tones of winter's plumage. But here were ruddy turnstones and red knots flashing their robin-red markings—red knots were once called "robin snipes"—as advertisements that breeding season was nigh. Even the normally monotone semipalmated sandpipers, stretched in a long line almost as far as the eye could see, displayed unusually bold, crisp markings. But no crabs were in sight; obviously they had arrived on the high tide to lay their eggs and then retreated again into the sea.

In search of crabs, I moved onto nearby Thompson's Beach. "Beach" might seem a misnomer, for the shore is nowhere to be seen. Sea level rise due to climate change has overwhelmed this former beachfront town, as it has many others along Delaware Bay, so that only pilings remain today, and one enters a pool-table-flat expanse of marshlands subdivided by tidal creeks. Thompson's Beach is now owned by one of New Jersey's major electric utilities, which has overseen the largest salt marsh reclamation project in the world, returning some 800 hectares (2,000 acres) of once-dyked salt marsh hay farms to the sea.

This morning, the results of this habitat transformation were on full display. A heavy morning mist hung over the marsh, forming a gray backdrop against which the silhouettes of drowned birch trees seemed projected as if onto a movie screen. The scene before me was perhaps the most primeval and, at the same time, the most exhilarating that I have ever witnessed in the natural world. Thousands of horseshoe crabs—so named for their shape and

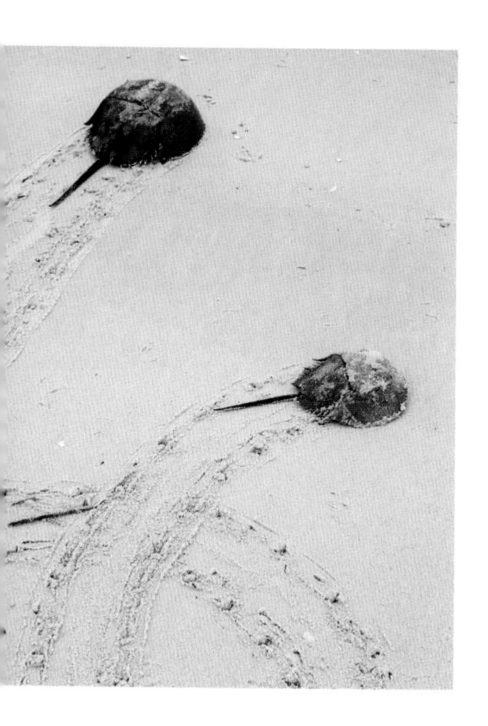

Horseshoe crabs come ashore to lay their eggs, as they have done for 400 million years.

size—were plowing heedlessly, blindly it seemed, onto the muddy banks of a tidal creek. Some were overturned, others were stranded among the dead stalks of salt marsh cordgrass, still others were pushing through the marsh muds in their instinctive drive to find an egg-laying site. There to greet them were thousands of shorebirds and laughing gulls and a small cadre of glossy ibises.

I had seen shorebirds feeding across the vast mudflats of the Bay of Fundy and seabirds diving relentlessly into schools of shrimplike krill, but never had I witnessed a feeding frenzy of such intensity. Red knots and dunlins were climbing over each other to get at the bounty of crab eggs. In their density and heedless behavior they seemed more like swarms of insects than birds. But it was perhaps the sound of this feeding frenzy—the unrestrained electric chattering of the shorebirds and the eerie chorus of laughing gulls—that stamped the spectacle as something primeval. Its pedigree reaches back through the ages, to a time before there was anyone to wonder at its mystery. This coming together of species in pursuit of food and the opportunity to mate has been aptly described by shorebird biologist J. Peter Myers as "sex and gluttony on Delaware Bay."

Horseshoe crabs are so-called living fossils, among the most primitive creatures still thriving on the planet. They arose in the Ordovician period, 400 million years ago, and have seen little reason to change their successful breeding behavior ever since. They have been repeating their spawning ritual since long before there was a Delaware Bay—or even a North American continent and an Atlantic Ocean as we know them today.

It is inaccurate to call them crabs at all, because they are arthropods more closely related to terrestrial spiders than to marine crustaceans. Their common name derives from the shape of the front portion of their body, or prosoma, which resembles a horseshoe. Underneath this impressive dome-shaped covering are five pairs of legs, four of which are tipped with pincers, which act as a food mill for grinding molluscs and worms and delivering them to the mouth at the base of the legs. The middle portion of the body, the opisthosoma, has a protective serrated edge, and underneath are housed so-called book gills, the breathing organ. The crab sports a tail-spike, or telson, which it uses for righting itself if it is flipped over by waves while spawning, and which native Americans employed as a spear point.

Horseshoe crabs reach sexual maturity at nine to ten years, when they make their spawning migration from deep waters of the bays or continental shelf to the beaches. They time their coming ashore to the highest lunar tides in May and June, at the new and full moons, to reach the higher intertidal portions of the beach. Often they hit the beaches on the night tides or in the low light of dawn or dusk, and they seek out beaches where there is protection from heavy surf.

Males arrive first and attach themselves to the female by locking onto her spiky opisthosoma with a pair of "boxing glove" pincers. Dragging the male—or sometimes more than one—the female works her way up the beach and excavates a nest 15 to 20 centimeters (6 to 8 inches) deep, where she deposits approximately four thousand tapioca-sized, greenish-gray eggs. When she turns to make her way back to the sea, the male fertilizes the eggs externally as he passes over the nest. The female will return to the beaches to spawn up to twenty times, depositing as many as eighty thousand eggs in a season. The return trips often churn up eggs that already have been laid, thus making them more readily available to the shorebirds. Horseshoe crabs live to be sixteen to seventeen years old. Their longevity, combined with the fact that they lay such an abundance of eggs, ensures their survival—provided they are not harvested too heavily, as has unfortunately been the case in recent history.

For a century, beginning in the early 1800s, crabs were harvested for fertilizer. They were first dried, then ground up in factories and spread on farmers' fields. The crabs were also fed to hogs and their eggs scooped up for chicken feed. Inevitably, the crab population collapsed in the early 20th century, and with it the unsustainable fertilizer industry. At the same time, shorebird populations had been drastically reduced by relentless gunning and market hunting.

Their exploitation of the crab eggs, which individually supply a minuscule caloric benefit, depends on a superabundance of eggs and therefore of crabs. Because shorebirds are both long-lived and opportunistic in their feeding habits, those that survived this onslaught on themselves and their prey most likely went elsewhere, perhaps utilizing the mudflats and sounds of New Jersey, which still attract as many as twenty species of shorebirds in spring. These include the short-billed dowitcher, dunlin, yellowlegs, black-bellied plover, semipalmated plover, and semipalmated sandpiper.

The crab harvest was revived in the 1980s, this time as a lucrative source of bait for conch and eel fishers. Again, horseshoe crab numbers collapsed by as much as 75 percent, and with them the number of the East Coast *rufa* race of red knots, which have declined by more than 85 percent on their spring migration to Delaware Bay—plummeting from 100,000 in the early 1980s to as few as 15,000 in 2005—and by 64 percent on their wintering grounds near Tierra del Fuego.

Shorebirds can accumulate fat quickly to fuel their remarkable intercontinental flights. In the 1980s, when crab numbers had rebounded to healthy levels, the red knots were gaining 8 grams of fat per day, but by 2002 they were gaining a quarter as much. The accumulation of fat is critical for the red knots to complete the last leg of their epic migration to their breeding grounds in Canada's High Arctic. To do so, they need to gain 6.5 grams per day for twelve to fourteen days, which means consuming approximately eighteen thousand horseshoe crab eggs every day. Otherwise, they will arrive in the Arctic without sufficient reserves to lay eggs and rear their young, or they will simply fall out of the sky from starvation en route. Those that do complete the journey often arrive before the weather has warmed enough to provide a reliable food supply, and so they must subsist on their stored fat reserves. Either way, accumulating enough fat is critical not only to reproductive success but to survival.

Delaware Bay has the largest concentration of spawning horseshoe crabs along the Atlantic coast. It is no coincidence that, historically, 90 percent of the Western Hemisphere's population of red knots stopped at Delaware Bay. Feeding sites such as Delaware Bay are critical links in a chain that allows shorebirds to complete their life cycle—which, for red knots, spans some 25,000 kilometers (15,500 miles) from wintering grounds in the Southern Hemisphere to breeding grounds in the Northern Hemisphere. Shorebirds follow what Brian Harrington of the Manomet Observatory for Conservation Sciences, in Massachusetts, has described as a "moveable feast." They concentrate at far-flung sites where food is uniquely abundant to fuel the next leg of their journey. This tendency for significant proportions of a whole population to concentrate at a single site makes shorebirds particularly vulnerable, since it breaks the normal link between the abundance of a species and its immunity to extinction.

*Migrant Trap at Land's End*

The Cape May Peninsula forms the northern shore of Delaware Bay and is a concentration point for migrating birds, especially in the fall. Vast numbers of songbirds stop over here on their southerly migrations, pausing before making the flight over the mouth of Delaware Bay to Cape Henlopen on the other side. In 2001, it is estimated that 1.25 million American robins, 300,000 red-winged blackbirds, 75,000 house finches, 2,500 eastern bluebirds, 2,000 rusty blackbirds, as well as thousands of sparrows, juncos, and yellow-rumped warblers converged on Cape May.

These birds were following what in the past has been referred to as the "Atlantic flyway," a term that is now debated if not defunct in ornithological circles. Birds, and other aerial migrants like butterflies and dragonflies (millions of monarch butterflies and dragonflies join this aerial parade), may follow certain lengthy geographic features such as mountain ranges or rivers. Migrants are attracted to these so-called leading lines for the advantages they provide, such as updrafts or ample food supply. By contrast, the Atlantic coast is thought to act as a diversion line along which migrants such as passerines and hawks concentrate because they want to avoid flying over the waters of Delaware Bay and the Atlantic beyond. Shorebirds fly great distances over

Dragonflies (left) join millions of birds that use the leading line of the Atlantic coast as a migratory route; the brackish marshes of Delaware and Chesapeake Bays are also home to amphibians and reptiles such as this rare bog turtle (right).

Delaware Bay is a very important wintering area for waterfowl such as these long-tailed, or oldsquaw, ducks.

the open Atlantic, but most birds, from hawks to songbirds, avoid flying over water if they can and so follow the coastline south. It is thought that the predominant northwest winds of autumn push most birds east of the Rocky Mountains toward the east coast, where they encounter the forbidding Atlantic and then head south along the coast, describing a dogleg migration.

For some species—especially wading birds and waterfowl—the coast may in fact function as a leading line that offers ample and familiar food supplies. And even for some songbirds the rich mosaic of coastal habitats may also be attractive, and in turn, the presence of these prey species acts as a magnet for raptors. However, the Cape May Peninsula of Delaware Bay and the Cape Charles Peninsula, which forms the north shore of Chesapeake Bay, become bottlenecks where this aerial river of southerly flowing birds slows and pauses before moving on. Land runs out at Cape May Point and birds must decide to fly over water, a distance of 17.5 kilometers (11.5 miles), or go around the bay, a diversion that, though less fraught with immediate risk, is nevertheless energetically costly. Cape May acts as a classic migrant trap, or bottleneck, where birds stop to refuel and to determine their migratory strategy.

The region around Cape May offers a diversity of habitat types, from oceanic waters to bays to salt marshes, and a little way inland, brackish marshes, freshwater lakes and ponds, and woodlands and fields. Birds pile up here before moving on, in the interim replenishing their depleted fuel supply. But many species also find this point, roughly midway between the far north and the deep south, a good place to stop and spend the winter, and large numbers of waterfowl pass the cold months at the mouth of Delaware Bay and in the many back bays along the New Jersey coast. Waters usually remain open in this marine-moderated region, making it a stronghold for American black ducks as well as brant. New Jersey hosts about 70 percent of the 140,000 brant that are estimated to comprise the Atlantic population. They arrive in mid-October and are among the last waterfowl to leave, fattening on sea lettuce in New Jersey's back bays for their long flight to James Bay, where they fuel up for the last leg of their journey to breeding grounds around northern Hudson Bay and Baffin Island. Greater scaup, red-breasted merganser, and bufflehead also flock to the back bay lagoons in the lee of the barrier islands for the winter.

But perhaps the most impressive gathering of waterfowl occurs off Cape May Point in February and March, when tens of thousands of scoters—all three species: surf, black, and white-winged—gather to feed on the rich invertebrate offerings in these offshore waters. The flocks, whose whistling enlivens the spring soundscape around Delaware Bay, are mostly males and include a very high percentage of the black and surf scoter population of the Atlantic Ocean. They are often joined by large flocks of long-tailed, or oldsquaw, ducks. Generally speaking, Delaware Bay and the waters surrounding the mouth are recognized as having the largest winter waterfowl populations in the Mid-Atlantic region. These same estuarine-enriched waters are thought to be the main wintering area for common loons, which feed on small flounders, sculpins, and crabs. They are joined by red-throated loons, which also stage there in spring, before moving northward.

*Where Freshwaters and Salt Waters Mix*

Captain John Smith was the first European to record the riches of Chesapeake Bay, in the journal of his 1608 voyage. Returning from a trip up the Potomac, where he had investigated a supposed silver mine that "proved of no value," he marveled at the natural capital of the Bay:

> ... a few Bevers, Otters, Beares, Martins and minkes we found, and in divers places that aboundance of fish, lying so thicke with their heads above the water, as for want of nets (our barge driving amongst them) we attempted to catch them with a frying pan; but we found it a bad instrument to catch fish with: neither better fish, more plenty, nor more variety for smal[l] fish, had any of us ever seene in any place so swimming in the water, but they are not to be caught with frying pans ...

The Savannah sparrow is among the maritime grassland birds in decline due to threats to its rare coastal habitat.

## MARITIME GRASSLANDS

MARITIME GRASSLANDS are rare coastal habitats restricted to the moraines that formed at the limit of the glaciers on Long Island and other southern New England sites such as Block Island, Martha's Vineyard, Nantucket, and Cape Cod. These "edge-of-the-ice communities" are found on the sandy soils that were deposited by the retreating Wisconsinan glacier and today are influenced by a maritime climate characterized by moderate temperatures, a long frost-free season, ocean winds, and salt spray. Also, much like the grassland communities of the Midwestern prairie, they seem in the past to have depended for their maintenance upon periodic fires. Bunch-forming grasses such as little bluestem, common hair grass, and poverty grass dominate these marine-influenced grasslands.

Sandplain grasslands, such as the Hempstead Plains on Long Island, have formed beyond the influence of offshore winds and salt spray. In the past, these grasslands were dominated by prairie-type grasses, including big bluestem, broom sedge, Indian grass, and switch grass; they also evolved with and were maintained by fire. They continue to provide important breeding and wintering areas to a number of declining grassland birds, including grasshopper sparrow, upland sandpiper, eastern meadowlark, and Savannah sparrow. Sandplain grasslands were also the refuge of the extinct heath hen—the last individual was seen on Martha's Vineyard in 1932.

Although Chesapeake's natural capital has been much depleted in the last four hundred years, the bay is still renowned for its historical productivity and present potential.

Chesapeake Bay is the largest estuary in the United States, stretching for 333 kilometers (200 miles) from Havre de Grace, Maryland, to Norfolk, Virginia, varying in width from a mere 5.7 kilometers (3.5 miles) near Aberdeen, Maryland, to 58.3 kilometers (36 miles) at its widest point, near the mouth of the Potomac River. More impressive than the dimensions of the bay itself is the size of the drainage basin, at 166,512 square kilometers (64,000 square miles). More than fifty rivers and thousands of streams fan out to form the bay's watershed, which includes parts of New York, Pennsylvania, West Virginia, Delaware, Maryland, and Virginia. Arising in the Appalachian Mountains, farther west, these freshwaters tumble from the fall line that forms the boundary between the Piedmont Plateau and the Atlantic Coastal Plain. The three main tributaries of the bay are the Susquehanna, Potomac, and James Rivers, with the great Susquehanna accounting for fully half of the freshwater input Altogether, the three rivers account for 70 to 80 percent of the bay's freshwater budget. They carry with them sediments and nutrients which, in part, are responsible for the historical productivity of the bay. Increasingly, they also carry a cocktail of contaminants from the industrialized and highly urbanized areas along their extensive reaches, where an estimated 16 million people work and live.

The Susquehanna has been called the mother river of the bay. During the last glaciation, the ancestral Susquehanna carved a valley through the coastal plain and onto the continental shelf, which, with so much water tied up in the glaciers, was exposed at the time. As the glaciers receded and began to give up their waters, the sea level rose and began rapidly drowning the river valleys. Seawater reached the area of Baltimore by nine thousand years ago and by three thousand years ago, the Chesapeake as we know it today had been created.

In addition to its impressive size—the most notable notch in the eastern seaboard of the United States—and the even more impressive area of its drainage basin, a three-dimensional view of the bay reveals another distinguishing if surprising feature: it is very shallow, averaging only 7 meters (23 feet) deep. This has profound implications for its legendary productivity. Estuaries are normally five times as productive as the open ocean, but the Chesapeake, with only about one-tenth the volume of water compared to most of the other major bays in the world, is more than a hundred times as productive.

The classic bay circulation is characterized by lighter freshwater moving across the top of the bay toward the sea, while the heavier salt water moves under this lighter layer toward the head of the bay. Winds frequently turn these layers over, and flushes of freshwater can also upset this classic circulatory pattern, turning it topsy-turvy. This constant mixing of salt waters and freshwaters means that the bay's dominant species must be able to tolerate wild swings in salinity. The salinity of the water affects all aspects of the bay's ecology, influencing spawning and nursery areas for crabs and fish, for example, and determining what plants grow where. Year-round residents, such as blue crabs, oysters, striped bass, and blue herons, must be adaptable to these ever-changing conditions.

Despite such vicissitudes, the turnover of waters in the bay does have a seasonal rhythm. In spring, the rivers bring a fresh supply of nutrients and oxygen, triggering a profusion of phytoplankton growth. The phytoplankon, in turn, attract fish and other consumers but eventually deplete the oxygen in these waters. In the fall, when river levels are low, tidal waters push farther into the bay and bring with them fresh supplies of oxygen. At the same time, the surface waters are cooling and therefore begin to sink, while the oxygenated salt waters now rise. In this way, the oxygen, and with it phytoplankton and zooplankton, are redistributed throughout the water column.

*following spread*: The Mid-Atlantic coastal region provides critical wetland habitat for a diversity of reptiles and amphibians, such as this bull frog.

The shallowness of the bay—10 percent of it is less than a meter (3 feet) deep and 20 percent is less than 2 meters (6 feet) deep—allows sunlight to penetrate much of the water column, fueling phytoplankton growth and supporting underwater meadows of sea grasses. More than a dozen varieties of sea grasses grow from the head to the mouth of the bay, depending on their varying tolerances to salinity, and provide food for waterfowl as well as nursery areas for shrimp, crabs, fish, oysters, and sea horses. The extensive tidal marshes that fringe the eastern shore of the bay—the so-called "Everglades of the North"—are the largest salt marshes in the Mid-Atlantic region. Here, the dominant cordgrasses—the *Spartinas*—rule, providing food or shelter for an array of marine creatures—from fish to oysters, clams to worms—and foraging areas for mammals such as mink, muskrat, nutria, and river otter. In these marshes and the adjacent shallow waters and tidal flats, eagles, osprey, and great blue herons also feed.

It takes six months for water from the Susquehanna, with its dissolved nutrients, to pass from the river to the ocean. During this period it is recycled several times, from top to bottom, a process abetted by the bay's shallowness. Nutrients taken up by the phytoplankton eventually sink to the bottom and

are recycled by the worms and clams, thus moving it through the food web. Underwater grasses absorb the phosphorus and nitrogen delivered to them by the river in spring and store it in their tissues until fall, when they begin to die back, thus releasing the nutrients to the system. They also bank nutrients in their roots, which are cropped by migrant swans and geese. Ultimately, this recycling of nutrients contributes to the bay's stability, a state of equilibrium that ecologists call homeostasis.

### *Keystone Crabs and Founding Fishes*

Of all the species associated with Chesapeake Bay's legendary productivity, the blue crab is the most iconic and, ecologically, one of the most important. Its scientific name, *Callinectes sapidus*, describes its physical appearance, behavior, and culinary qualities. *Callinectes* is Greek for "beautiful swimmer," and *sapidus* means "tasty" in Latin. Some may argue whether its common name, blue crab, is descriptive enough, for it is only the large claw and legs of the male that are truly blue. The blue crabs' tastiness, however, is beyond dispute, and they are very able swimmers. Blue crabs belong to the *Portunidae* family, and unlike most other crabs, portunids have a special pair of swimming legs allowing them to move in any direction, and even to hover in place like an underwater helicopter.

BLUE CRAB

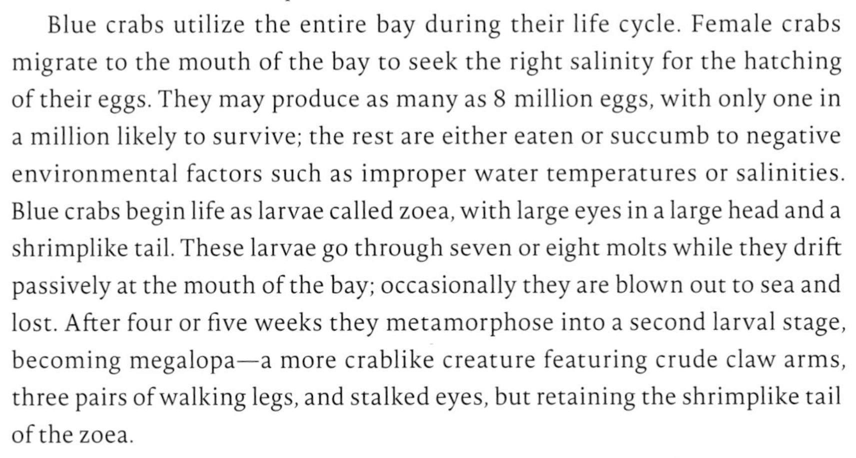

Blue crabs utilize the entire bay during their life cycle. Female crabs migrate to the mouth of the bay to seek the right salinity for the hatching of their eggs. They may produce as many as 8 million eggs, with only one in a million likely to survive; the rest are either eaten or succumb to negative environmental factors such as improper water temperatures or salinities. Blue crabs begin life as larvae called zoea, with large eyes in a large head and a shrimplike tail. These larvae go through seven or eight molts while they drift passively at the mouth of the bay; occasionally they are blown out to sea and lost. After four or five weeks they metamorphose into a second larval stage, becoming megalopa—a more crablike creature featuring crude claw arms, three pairs of walking legs, and stalked eyes, but retaining the shrimplike tail of the zoea.

After a week, the megalopa undergoes another metamorphosis into a recognizable juvenile crab capable of swimming and walking on the bottom of the bay. These tiny crabs now reverse their mother's migratory direction and begin working their way up the estuary toward its head. They keep close to the shore, taking shelter in the sea grasses to avoid predators, including mature blue crabs, which will cannibalize the young. It may take seven or eight months for them to migrate half the length of the bay, by which time it is late autumn. The cold waters then send them to the bottom of the bay, where they dig into the mud to hibernate. In spring, the new crabs resume their steady migration toward the head of the bay.

Crabs may live up to three years, though many are caught by the bay's traditional fishers, or "watermen," or succumb to natural mortality before then. Crabs regularly outgrow their hard shell, or exoskeleton, and must shed it, a process known as molting, which may occur as many as twenty-seven times during their lifetime. During the three-day period it takes for the shells to harden, the crabs are particularly vulnerable to predation. Molting, however, is critical to the crab's growth and survival, as well as to successful reproduction. When females become sexually mature and are about to molt, they release chemicals into the water that attract males.

Blue crabs have a remarkably elaborate mating ritual. Males raise themselves on the tips of their walking legs and wave their arms as an attention-getting ploy, at the same time kicking up sand with their swimming legs. Recent research indicates that the arm waving by the males may help to waft sexually attractive scents—pheromones—to the prospective mate. Receptive females respond favorably by rocking back and forth, then backing up under the raised male, who grasps his partner with his claws. The pair assumes what is known as the cradle position, with the male carrying the receptive female underneath him for one or two days until she is ready to molt. The pair separates briefly while the female undergoes her molt; then they resume their pas de deux. The male now gently overturns the female and inserts, face-to-face, his pleopods, containing the sperm, into the female's genital pores. The mating may take from five to twelve hours, during which the female expands her abdomen so that it folds around and over her partner's back to hold him in place.

After copulation, the pair resumes the cradle position for forty-eight hours. This behavior has two possible purposes: it may protect the female during the vulnerable post-molt stage, or the male may simply be protecting his own posterity by preventing the female from mating with another male. Now the female begins her migration toward the mouth of the bay, while the male remains in the brackish rivers and inlets. The female fertilizes her eggs during her migration but may hold back a sperm packet for nearly a year for a second fertilization, a strategy that may compensate for poor environmental conditions in any given year.

Blue crabs have been recognized as a keystone species in Chesapeake Bay, important as both predator and prey in the bay's food web. As predators, they feed at more than one level of the food web, and therefore their increase or decrease can have cascading effects on the rest of the ecosystem. It has been shown, for example, that blue crabs enhance salt marsh production by feeding upon marsh periwinkles, which at high densities can reduce salt marsh grasses by their grazing activities. The crabs' preferred prey is the hard shell clam, or quahog, the most common clam in the Chesapeake, but as omnivores they also prey on a variety of small fishes, marine worms, and other crustaceans, like shrimp, barnacles, mud crabs, and immature blue crabs, as well as plants. Juvenile crabs are in turn preyed upon by striped bass and other fishes, including drums, eels, catfish, cownose rays, and some sharks. As larvae they are consumed by filter feeders such as menhaden, bay anchovies, and even oysters. From this list of predator and prey, it is clear that blue crabs are connected to all levels of the food web, and so changes in their numbers have ripple effects throughout the system.

QUAHOG

IN 1936, RACHEL CARSON wrote: "Just as the sacred cod of Massachusetts is the accepted emblem of [that] state, so the shad may rightly be considered the piscatorial representative of the states bordering the Chesapeake." The famed *New Yorker* staff writer John McPhee went further, arguing that the American shad was "the founding fish" of the United States. In the fourth spring of the American Revolution, George Washington, himself a commercial shad fisherman, bivouacked his army on the banks of the Schuylkill River. Historians believe that the founding father knew exactly what he was doing, because the spring run of fat shad reputedly saved his men from starvation.

The American shad is the largest member of the herring family, ranging in weight from 1 to 4 kilograms (2 to 8 pounds) and, though bony, is justly famed for its culinary qualities, which earned it the species name *sapidissima*, or "most delicious." Herring and shad were the basis of the first commercial fisheries in the Chesapeake Bay, and on the Potomac River alone, six thousand fishermen harvested these two species in the 1830s. Large spawning populations once reached far into the hinterland of Chesapeake's waters, where they were also important food sources for the settlers of these regions. Shad migrated up the Susquehanna River as far as New York State and reached the foothills of the Blue Ridge Mountains via the mighty James River.

Much of the spawning territory on these and many other spawning rivers along the eastern seaboard was lost in the 19th century to dams. The Chesapeake Bay writer Tom Horton has observed that these dams created "the amputated bay," in which thousands of kilometers of spawning territory was cut off with predictable results—a collapse of shad stocks. It was not until the last decade of the 20th century that the situation was corrected with the building of more than a hundred fish ladders and the removal of nearly forty dams. As a result of these steps, in concert with an ambitious restocking program and a moratorium on commercial fishing of stocks as they migrate along the coastlines of Virginia and Maryland, the shad has made a strong comeback. Its smaller relative, the alewife, or river herring, has also benefitted from these measures, since it makes spawning runs on many of the same rivers as the shad.

AMERICAN SHAD

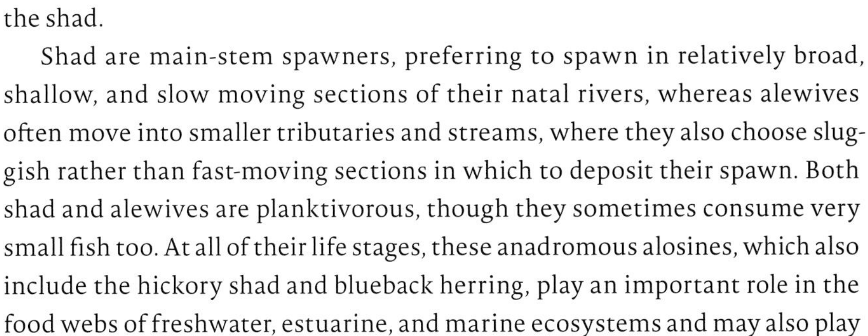

Shad are main-stem spawners, preferring to spawn in relatively broad, shallow, and slow moving sections of their natal rivers, whereas alewives often move into smaller tributaries and streams, where they also choose sluggish rather than fast-moving sections in which to deposit their spawn. Both shad and alewives are planktivorous, though they sometimes consume very small fish too. At all of their life stages, these anadromous alosines, which also include the hickory shad and blueback herring, play an important role in the food webs of freshwater, estuarine, and marine ecosystems and may also play a significant role in transferring nutrients from the marine system to freshwater rivers.

Another fish, striped bass, has also long been identified with the Chesapeake Bay. Like shad, striped bass are an anadromous species that spawn in fresh or brackish water and return to ocean waters. They are found year round in the bay, which serves as the nursery for juvenile striped bass on the Atlantic coast. Seventy to 90 percent of the Atlantic population spawn and spend their first years of life in the bay, where they feed on a variety of small crustaceans, including blue crabs. In 2000, a study estimated that striped bass, or rockfish as they are better known in the Chesapeake, were eating approximately 75 million small blue crabs that take refuge in the underwater grass beds in the middle of the bay. Large as this number appears to be, it is only 5 percent of the total population of young crabs taking shelter there. Rockfish themselves were severely depleted by overfishing and were on the brink of collapse when a moratorium was declared in the early 1980s. Their numbers rebounded to historical levels by the end of the decade.

Like blue crabs, striped bass are voracious predators, preying on a variety of marine life. Perhaps their most important prey is menhaden, another member of the herring family, and as such a filter feeder, straining plankton through its gill rakers. It is arguably the most important fish species in the

Chesapeake Bay serves as a nursery for striped bass, or rockfish, for the entire Atlantic coast population.

bay ecosystem. As the biologist W.K. Brooks noted in 1903, "All our best and most valued food fishes are only menhaden in another shape." The decline in the menhaden population in Chesapeake Bay has led some scientists to suggest that, as a result, striped bass may be suffering from malnutrition, which in turn has led to high levels of a chronic wasting disease known as mycobacteriosis. If so, this condition serves as another example of how the abundance and health of one species may affect the welfare of another in this closely interconnected estuarine ecosystem.

*Ecosystem Engineers*

The eastern oyster occurs in coastal waters from the Gulf of St. Lawrence to Argentina, but like the blue crab and shad it has played a historically important role in Chesapeake Bay. Chesapeake is an Algonquian word meaning "great shellfish bay." At White Oak Point, on the lower Potomac River, an oyster shell midden is testament to four thousand years of continuous harvests. Oysters became the source of a major fishery in colonial times in areas such as Chesapeake Bay, Narragansett Bay, and Long Island Sound. In the 1880s, at the height of local production, the Chesapeake was the greatest oyster-producing region in the world. Predictably, like the blue crab and shad, the eastern oyster suffered from severe exploitation, and its numbers

Chesapeake is an Algonquian word meaning "great shellfish bay," named for the eastern oyster once so abundant there.

have been reduced a hundredfold in the last 150 years. More recently, parasitic diseases have further ravaged surviving oyster beds. Although this loss has had a direct economic impact, its effect on the ecosystem is perhaps more profound.

The reproduction of the eastern oyster is heavily influenced by temperature. The ideal temperature is between 18° and 25°C (64° and 77°F), one that occurs in the Mid-Atlantic region from late May to September. Males are stimulated to spawn by temperature as well as food supply, and tend to spawn first. The presence of spawn in the water then stimulates the females to release eggs, which are fertilized externally. The larval stage lasts two to three weeks, again depending on food availability and temperature. During this time the larvae migrate vertically, tending to concentrate near the bottom during the outgoing tide and rising in the water column during the incoming tide, a behavior that increases their chances of being retained in the estuary. Eventually the larvae must settle on a clean, solid surface, preferably the shells of other oysters. This gregarious behavior enhances the larvae's ability to successfully reproduce and provides protection from predators and other physical stresses. It also results in the building of oyster reefs, which, in the past, carpeted acres of shallow water habitats and sometimes grew to the truly impressive height of 4 meters (13 feet).

## COASTAL PILGRIMS

Dunes on the move bury pitch pine near Provincetown, Massachusetts; (facing page) a shallow scrape provides scant protection for eggs of the endangered piping plover.

HUMANS HAVE become a geological force, initiating and abetting the movement of barrier islands, albeit inadvertently. When the Pilgrims first came to the Province Lands of Cape Cod, Massachusetts, in 1620, they began to eliminate the maritime forest of bur oak and pitch pine for firewood and to use the beach grass–clad dunes as pastures for their cattle. The result, we now know, was predictable: left exposed to the winds, the dunes began to migrate toward Provincetown with alarming rapidity and by the late 18th century were burying parts of the village. It was only through the planting of beach grass and trees that the advance was slowed, but in places today the dune's advance can still be seen slowly burying the protective forest and creeping across roads.

Although we have had ample time to learn the lessons of interfering with the natural evolution of barrier islands, many islands have been built over with condominiums, hotels, summer homes, and shopping centers, completely obliterating habitat. But even where the dunes have been protected from the bulldozer, intensive recreational use has continued to degrade and often destroy the mosaic of

habitats on the islands. The careless operation of motorized recreational vehicles, and even intensive pedestrian traffic, has caused damage to the beach grasses that hold dunes together. As a result, rapid erosion can occur rather than the more gradual, and natural, process of island migration. Although the beach itself is perhaps the habitat least vulnerable to human use, even here the use of vehicles and the throngs of beachgoers on foot can disturb nesting adults and imperil the young of beach-nesting species such as terns, plovers, and skimmers. The ridges created by tire tracks have also been implicated in preventing newly hatched turtles from making their perilous first migration from beach nests to the sea, as with loggerhead turtles, which make nests and lay their eggs on some barrier islands from North Carolina south to Florida.

The damage is more deliberate and extensive when salt marshes—the most productive of habitats—are drained as an attempt at mosquito control. Studies at Assateague Island, Maryland-Virginia, showed that ghost crab numbers were significantly lower on beaches subject to off-highway-vehicle traffic than on wild or pedestrian-only beaches. Mole crabs also are severely affected by vehicular traffic, even though OHVs rarely enter the swash zone. Since ghost crabs and mole crabs occupy a midpoint in the food web, breaking down and digesting detritus and serving as food for fish and birds, their loss has ramifications throughout the food web.

These remarkable living structures have earned the eastern oyster praise as an ecosystem engineer. Reefs perform a number of important ecological functions, serving as predator refuges as well as feeding and nesting sites for a wide variety of species. With their rugged and ragged architecture, reefs contribute to habitat complexity and thus enhance productivity and biodiversity. More than 300 species have been found living and growing in the reefs.

STRIPED BLENNY

Chesapeake Bay oyster reefs have among the highest densities of fish ever recorded outside of tropical coral reefs and are home to a greater diversity of species than the unstructured soft-bottom habitat nearby. Some fifty-seven species of fin fish have been spotted on oyster reefs, including white perch, black sea bass, striped bass, and Atlantic silverside. The species that frequent oyster reefs fall into three groups: reef residents, such as the blennies and gobies, which use it as a primary habitat; so-called facultative residents, which are often found in and around reefs; and transient species, which are generally more far ranging but tend to forage on or near reefs. Some residents, like the naked goby and striped blenny, become prey for transient species such as striped bass. Generally, however, reefs serve as nursery habitats, necessary for the breeding, feeding, and growth of gobies, blennies, skilletfish, and oyster toad fish. In winter, diving ducks and loons feed on and over the reefs.

Large oyster reefs perform other ecological services, such as preventing shoreline erosion. But perhaps their most important role is as a filtration system. As filter feeders, oysters pump 80 to 100 liters (21 to 26 gallons) of water through their bodies daily. In the process, they glean phytoplankton as food and remove particulate matter from the water column, which they transport to the bottom. The result is clearer water, which allows more light to penetrate, enhancing the growth of sea grasses and bottom-dwelling algae. As we have already seen, sea grass beds serve as important nursery habitats for blue crabs and striped bass. It is difficult to estimate the overall positive effect of oyster reefs on the ecosystem, but some idea of the scale of the impact can be deduced from this fact: at one time, oysters could recycle the entire water volume of the bay in 3.3 days, a process that now takes 325 days.

Clearer water means more growth of submerged underwater vegetation, which depends upon water clarity for photosynthesis. Some sixteen such sea grasses grow in the Chesapeake, each occupying a different part of the bay, from the mouth to the head, depending upon its tolerance to varying salinities. Wild celery grows in freshwater; sago, pondweed, redhead grass, and widgeon grass in the brackish waters of the mid-bay; and eelgrass at the mouth of the bay, where waters are essentially salt. Declining water quality due to sedimentation and algal growth has reduced visibility to 1 meter in many areas of the bay, with the consequence that sea grass beds have declined in extent by as much as 90 percent since the 1970s.

Like oyster beds, sea grass beds act like bio- or ecological engineers, providing multiple environmental services. The grasses act as shock absorbers for wave energies, thus reducing shoreline erosion. Their undulating blades also trap sediments and consolidate them around their roots. In addition, the grasses sop up nutrients such as phosphorous and nitrogen, which pour into the bay from a host of sources, ranging from agricultural runoff to urban sewage. Their ability to absorb these nutrients (which in excess act as pollutants) prevents algal bloom in the spring. In the fall, when the grasses die back, they gradually release the nutrients into the bay, again mitigating their potentially negative effects on water quality.

The sea grass beds also greatly enhance the physical complexity of the shoreline environment, and as a result become a hiding, feeding, and breeding place for a diversity of organisms. A handful of healthy sea grass is alive with shrimp, little fishes and crabs, sea horses, and diamond back terrapins. Eelgrass blades also harbor a rich epiphytic community—microalgae, hydroids, bryozoans, sponges, and barnacles—which use the blades as support. Like oyster beds, sea grass beds serve as important nursery habitats for shellfish and finfish, and eelgrass beds in particular are especially important to juvenile blue crabs, which use them to avoid predators. Sea grasses are also an important source of food for resident and migratory species. Bay scallops depend upon eelgrass beds for their detritus-based diet and also use it as a nursery area, attaching themselves to the blades until they are large enough to survive on the sea bottom. Many of the more than a million waterfowl that depend upon the bay exploit sea grasses—redhead grass is likely named for redhead ducks that prefer it as fodder.

Collapses of this grass due to poor water quality caused a marked decline in the numbers of redhead duck utilizing the bay, from eighty thousand a half century ago, to fewer than ten thousand today. But overall, underwater grasses throughout the Chesapeake are making a comeback as measures to improve water quality take effect, an upswing which bodes well for the health of the bay.

NARRAGANSETT BAY AND Long Island Sound are remnants of large river basins at the margin of the glacier's most southerly advance, where today the glacial debris forms cobble beaches along their shores. These protected bays harbor relatively simple intertidal communities compared with the rocky shores north of Cape Cod, which acts as a barrier to the richer boreal fauna found there.

Cobble beaches, composed of small stones, have a high intertidal zone dominated by the cordgrass *Spartina alterniflora*, which acts as a buffer against wave shock and erosion. Cobbles nestled among the root mat of the cordgrass are often encrusted with barnacles and mussels, and typical salt-loving plants, halophytes such as goosefoot and glasswort, establish behind the cordgrass stands, which trap the seeds of these annual plants and prevent them from being washed away. In so doing, cordgrass acts as a bioengineer, over time transforming cobble beach habitats into high marsh plant communities.

Barnacles and fleshy algal crusts cover the cobblestones of the middle intertidal zone. But the most conspicuous organism is the common European periwinkle, which occurs in astronomical numbers, six hundred to one thousand per square meter, in Narragansett Bay and Long Island Sound. This herbivorous snail is an invasive species that appeared in Nova Scotia in the mid-19th century and moved south, reaching Chesapeake Bay by the 1950s. It is a habitat generalist now found in all northeast shoreline habitats. In salt marshes it has replaced the native mud snail, and on estuarine cobble beaches it severely restricts the growth of green algae. Such an algal canopy would lead to sediment accumulation and the establishment of tube-building organisms, such as worms and crustaceans, typical of soft-sediment communities. The periwinkle, however, is restricted to the intertidal zone—perhaps by predatory whelks or crabs—allowing an algal canopy of Irish moss, dead man's fingers, and sea lettuce to develop in the subtidal zone.

Humans are undoubtedly the most effective bioengineers, although our impacts on ecosystems are often inadvertent. The Hudson River estuary and New York Harbor are among the most polluted estuarine environments in the United States, repositories of industrial wastes, such as heavy metals and toxic organic compounds, and untreated sewage. Commercial fisheries in these waters shut down in 1976 because of PCB contamination. Pollution has been dramatically reduced and a number of native fish species have rebounded in recent decades, including American shad, striped bass, shortnose sturgeon, and the Hudson's behemoth, the Atlantic sturgeon, which can reach 4 meters (13 feet) and 360 kilograms (800 pounds).

The area between Boston and Washington, D.C., is one of the most densely populated urban areas in the world, home to tens of millions of people whose cities, towns, and suburbs meld into a single urban entity. Even so, the shore and offshore environment near this megalopolis remain surprisingly vital. As

A resolute royal tern faces into the wind; its survival depends upon limited breeding sites along the Mid-Atlantic shore.

we have seen, it is host to a large number of marine birds, both breeders and nonbreeders. A number of these local breeders, such as gull-billed tern, roseate tern, and Forster's tern, nest at a very limited number of sites, as do an important proportion of the western Atlantic population of royal terns. The Chesapeake feeds impressive numbers of bald eagles and is home to the largest population of osprey—some two thousand nesting pairs—in the world. The invertebrates and plants of both the Chesapeake and Delaware also support large numbers of waterfowl; tundra swans, Canada geese, greater snow geese, and a variety of ducks, including pintails, canvasbacks, eiders, and ruddy ducks, winter here.

Fish species that belong to the cold-temperate, or boreal, province north of Cape Cod and the warm-temperate province south of Cape Hatteras also meet and mingle in these waters on a seasonal basis. Most of the fishes found here, in fact, are migrants, and a few, like scup, summer flounder, longfin squid, and butterfish, are endemic to the region. When waters begin to warm in the spring, there is large influx of southern species such as drums, bluefish, and jacks into the Middle Atlantic area between the capes. At the same time, cold-water species such as cod, Atlantic herring, Atlantic mackerel, spiny dogfish, and shad, which had migrated to the Mid-Atlantic during the winter, now begin moving north. This north-south changing of the guard reverses in autumn, when water temperatures again begin to drop.

Offshore, sea turtles—loggerhead, green, hawksbill, and Kemp's ridley—ply these waters in summer, while bottlenose dolphins are often seen cavorting near the coast, from Long Island to Cape Hatteras.

# 5

# TIDES OF LIFE

## *The Gulf of Maine and the Bay of Fundy*

ON APRIL 24, 1895, Captain Joshua Slocum set out from Boston on the first solo circumnavigation of the globe, with "the waves dancing joyously across Massachusetts Bay." He soon put Marblehead astern and set his sails for "Gloucester's fine harbour," where he provisioned his little sloop, *Spray*, for the long voyage ahead. But first Slocum wanted to visit his birthplace, Westport, on Brier Island, near the mouth of the Bay of Fundy. Along the way, he anchored to handline "three cod and two haddocks, one hake, and, best of all, a small halibut." Then heading east, he passed the many islands of Maine's coast, one of which he dubbed "Island of Frogs," for the million-strong chorus of spring peepers that echoed across the water. The light of Gannet Rock welcomed him into home waters, where he passed over the "worst-tide race in the Bay of Fundy," which, he confessed, gave him a "terrible thrashing."

In this first leg of his great adventure, chronicled in the classic *Sailing Alone Around the World*, Slocum traversed nearly the entire breadth of the Gulf of Maine before entering the Bay of Fundy. We now recognize that these two linked bodies of water form a single oceanographic unit that stretches from the dramatic dunes of Cape Cod, along the rocky, island-studded coast of Maine, to the great expanses of mudflat and salt marsh at the head of the Bay of Fundy, home of the highest tides in the world.

‹ A lone fisher greets the dawn in the rich waters where the Gulf of Maine and the Bay of Fundy meet.

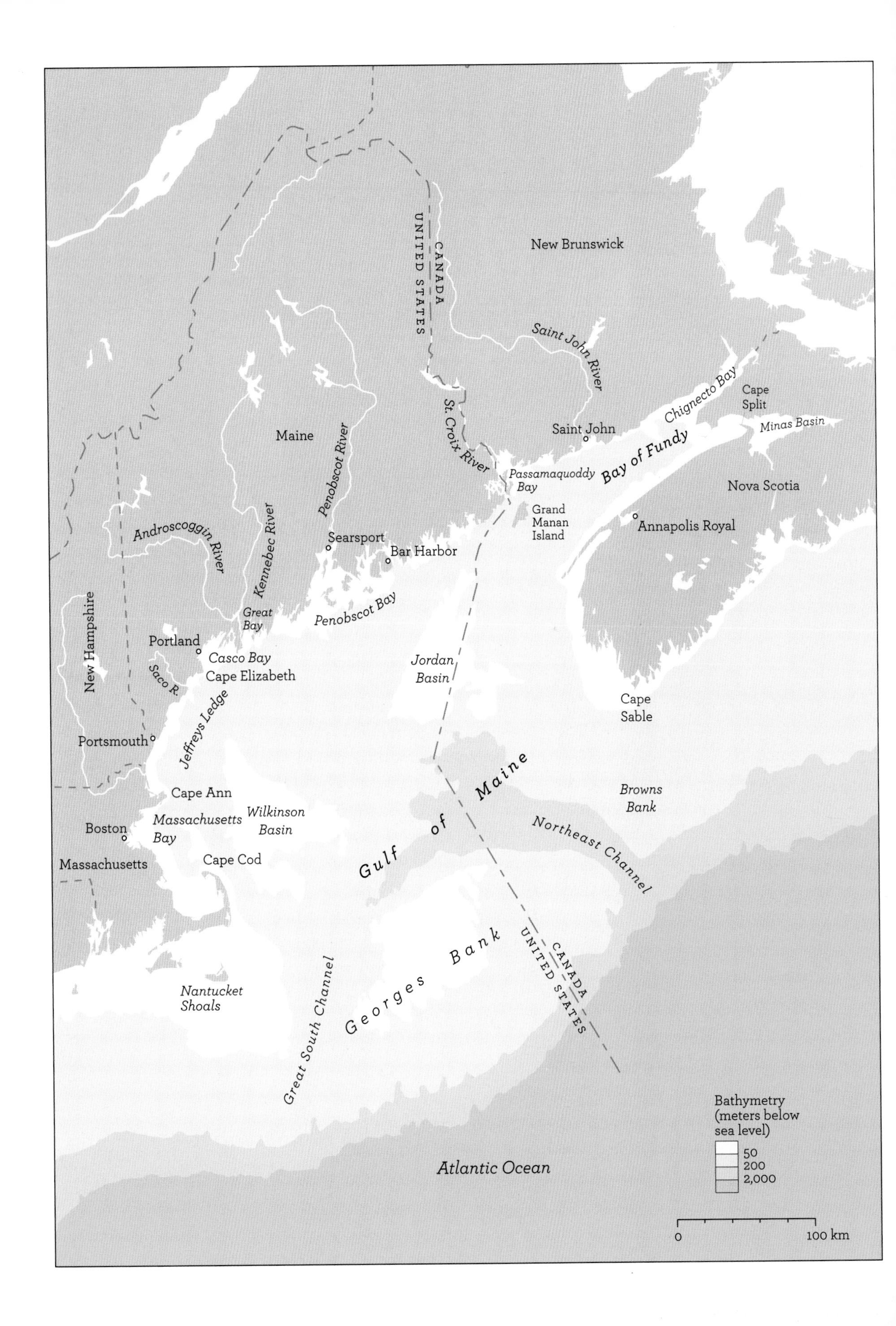

CANADA
UNITED STATES
New Brunswick
Saint John River
St. Croix River
Saint John
Chignecto Bay
Cape Split
Minas Basin
Bay of Fundy
Passamaquoddy Bay
Nova Scotia
Grand Manan Island
Annapolis Royal
Maine
Penobscot River
Kennebec River
Androscoggin River
Searsport
Bar Harbor
Penobscot Bay
Great Bay
New Hampshire
Portland
Casco Bay
Cape Elizabeth
Saco R.
Jeffreys Ledge
Jordan Basin
Cape Sable
Portsmouth
Cape Ann
Gulf of Maine
Browns Bank
Boston
Massachusetts Bay
Wilkinson Basin
Northeast Channel
Massachusetts
Cape Cod
Georges Bank
CANADA
UNITED STATES
Nantucket Shoals
Great South Channel
Bathymetry (meters below sea level)
50
200
2,000
Atlantic Ocean
0
100 km

Yachts rest at anchor at Marblehead, Massachusetts, while cormorants enjoy a pierside perch.

*facing page:*
MAP: Gulf of Maine and Bay of Fundy

Although apparently open to the sea, the Gulf of Maine is a marginal sea unto itself, largely cut off from the Northwest Atlantic by two fish-rich banks, Georges, which extends like a giant thumb off the outstretched arm of Cape Cod, and Browns, which lies off the southwestern tip of Nova Scotia. In essence, these banks are great sand dunes that were formed as outwash from the retreating glaciers, and today they rise from the ocean bottom to within a few tens of meters of the wave tops.

Two currents dominate the water circulation within this marginal sea: a counterclockwise gyre in the Gulf and an adjacent clockwise gyre on Georges Bank. The Nova Scotia Current, which is a cold offshoot of the Labrador Current, enters the Gulf of Maine through the Northeast Channel and is promptly deflected to the north, into the Bay of Fundy, by the Earth's rotation (the Coriolis effect). It then describes a counterclockwise gyre, skirting, in turn, the coastlines of New Brunswick, Maine, New Hampshire, and Massachusetts, before the upturned arm of Cape Cod redirects its slow journey, which takes about three months to complete. At its southern boundary, the current encounters the northern edge of Georges Bank. Some of the water skirts the bank and continues on its counterclockwise journey, while the rest flows around the northeast peak, helping to create the second gyre, which circles the bank in a clockwise direction. Water that does not complete the circuit of the bank spills into the North Atlantic through the Northeast Channel, or it may be lost through the Great South Channel, near Nantucket Shoals, before flowing southward to the Mid-Atlantic.

This is a simplified picture of a very complex circulatory pattern that expresses itself in marked variations in temperature, salinity, and nutrient

Common dolphins dive through the rich waters of the Gulf of Maine, feeding ground for cetaceans large and small.

content of the water, depending on the season, location, and water depth. Critically, these currents, or gyres, together with tidal action, combine to produce conditions that are conducive to life in the sea.

*Garden in the Sea*

The Gulf of Maine has been called "a garden in the sea" for its high productivity. Most of the oceans of the world are, in fact, watery deserts, largely because of a lack of available nutrients to feed the growth of phytoplankton, the single-celled plants that are the foundation of biological productivity in the oceans. Besides a supply of nutrients, phytoplankton need sunlight to power the process of photosynthesis, whereby they convert the sun's energy into organic material and produce oxygen as a by-product.

These two ingredients—nutrients and sunlight—are amply satisfied within the Gulf of Maine. Nutrients are continuously supplied from the Atlantic by waters seeping through the Northeast Channel, which is 61 kilometers (38 miles) long and 35 kilometers (22 miles) wide. The gulf is also 5°C colder than water to the south of it, and these cold waters are critical for providing the right combination of temperature and salinity needed by cod, haddock, flounder, and other fishes. The relatively low salinity and temperature of the coastal gulf waters is due to the freshwater from rivers pouring directly into the gulf as well as the massive outflow of freshwater from the far away St. Lawrence River.

These cold waters hold more oxygen and carbon dioxide in solution than warmer water, amplifying the potential productivity of the gulf. And the nutrient load of the gulf is substantially supplemented by the freshwater—some 950 billion liters annually—from the sixty rivers flowing into it, of which the Saint John is the largest. This great drainage basin collects waters from three American states and three Canadian provinces: Massachusetts, New Hampshire, Maine, Nova Scotia, New Brunswick, and even a portion of Quebec. These freshwaters, flowing downhill from a web of streams, ponds, lakes, and rivers, enrich the marine ecosystem with a constant and critical supply of nutrients from the land. But it is the currents and tidal action that bring these elements together, blasting nutrients from the sea bottom into the sunlight, or photic zone, thus kick-starting the chain of events that creates the classic marine food web: phytoplankton, shrimp, herring, whales.

Phytoplankton, or algal blooms occur under special conditions, whenever there is an adequate supply of both nutrients and sunlight. In summer the surface waters of the gulf are warm, but the nutrients are located in its deep basins. As the amount of sunlight declines, however, the surface waters begin to cool, and as they do, they grow heavier and begin to sink. This process, known as convection overturn, has been compared to turning over a garden, bringing nutrients from the deeper layers toward the surface. When these nutrients meet the light, there is an explosion of growth—an algal bloom. This occurs in the spring and fall in the central basins of the gulf, with dramatic results. Wherever you find phytoplankton blooms you will also find fish, seabirds, and whales feeding.

The link between the primary producers and the higher marine life forms is the zooplankton, a mixed group of organisms that includes tiny crustaceans, worms, molluscs, and larval fish. The most dominant group is the copepods, small shrimplike crustaceans numbering 7,500 species globally and occurring on land and in freshwater as well as in marine habitats. Copepods are generally the dominant group in marine zooplankton communities, where they primarily graze on diatoms and other large phytoplankton by filtering them from the seawater. They can also be carnivorous, however, preying on protozoa and other smaller zooplankton, including larval fish. Although they are able to swim freely, they cannot make headway against a current. But

## THE *Calanus* COMMUNITY

*CALANUS FINMARCHICUS* is the most important copepod in the Gulf of Maine, and in winter is found throughout the Atlantic Ocean, from the Arctic south to Chesapeake Bay. Because of its relatively large size—more than 3 millimeters (0.1 inches) as an adult—it is the largest contributor to biomass in the zooplankton community in the gulf, even though it may not always be the most abundant species. Its presence is so dominant on Georges Bank that famed oceanographer Henry Bigelow described the Georges Bank ecosystem as a "*Calanus* community."

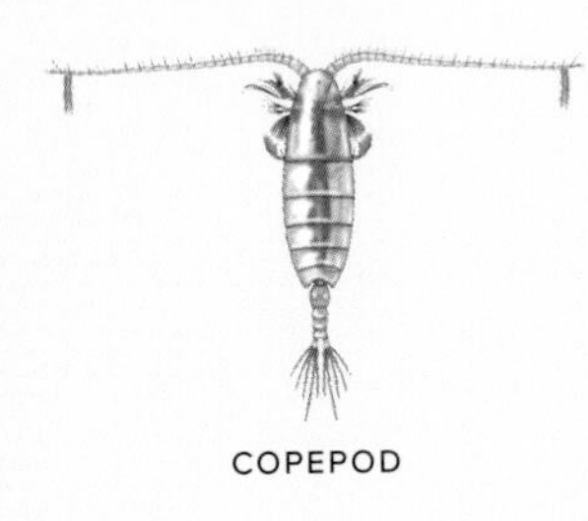

COPEPOD

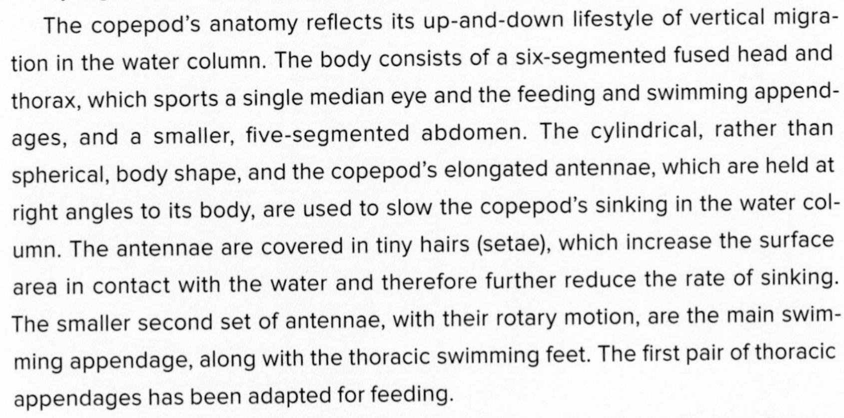

The copepod's anatomy reflects its up-and-down lifestyle of vertical migration in the water column. The body consists of a six-segmented fused head and thorax, which sports a single median eye and the feeding and swimming appendages, and a smaller, five-segmented abdomen. The cylindrical, rather than spherical, body shape, and the copepod's elongated antennae, which are held at right angles to its body, are used to slow the copepod's sinking in the water column. The antennae are covered in tiny hairs (setae), which increase the surface area in contact with the water and therefore further reduce the rate of sinking. The smaller second set of antennae, with their rotary motion, are the main swimming appendage, along with the thoracic swimming feet. The first pair of thoracic appendages has been adapted for feeding.

Copepods reproduce sexually. The male grasps the female with its antennae and, using its thoracic appendages, transfers a sperm package called a spermatophore to the female at the genital opening. The fertilized eggs are shed into the environment and develop through six stages. At each molt the organism undergoes metamorphosis, adding new appendages or modifying existing ones. During the fifth phase of its life cycle, *Calanus* enters a state of diapause, a period of delayed development and reduced metabolism. During this phase, at the onset of the warmest months of late summer, the copepods migrate to the deepest basins in the gulf, where they remain through the fall and winter. They undergo their final molt in the spring, emerging as adults and migrating vertically again to feed on the phytoplankton riches at the surface, and in turn to be fed on by predators higher in the food web, including fish, seabirds, and whales.

zooplankton undergo daily vertical migrations, a defensive behavior designed to prevent predators from seeing them. (Older animals—especially egg-bearing females—may also gain an energetic advantage by spending their days in deeper waters, because their rate of oxygen consumption drops so much.) During the day they concentrate in deeper, dimmer water, and at night they migrate into the upper water layers, where their phytoplankton food items are more abundant. Gulf of Maine fishermen call the reddish-brown adults "cayenne" when they appear in great swarms on the surface.

Although the Gulf of Maine as a whole is one of the most productive marine regions in the world, Georges Bank is particularly so, since it is shallow and has a ready supply of minerals and nutrients on its dunelike seabed. For its size, Georges may be the richest marine area in all of the temperate and high latitudes of the world's oceans. It is so productive that one Russian trawler captain described it in his memoirs as "an oceanic miracle."

Since the early part of the 20th century, there has been a concerted fishery throughout the year on Georges Bank, which has revealed the seasonality of fish movements. Traditionally, there was a winter fishery for fresh haddock and other mixed species, a spring halibut fishery that often began in late winter and extended into the summer months, a summer fishery for cod, and a swordfish fishery in June, July, and September. Since World War II, however, the fishery for the sedentary scallop has become dominant economically.

Georges's nurturing capacity is obvious even to the casual observer. I spent a week on a scallop dragger on the bank, and at any given time I might see whales feeding or shark or swordfish fins knifing the waves, and always hundreds of shearwaters (breeders from the Southern Hemisphere summering

This anemone (left) and sea cucumber (right) are part of the cornucopia of bottom-dwelling creatures common to the Gulf of Maine.

in the gulf during the austral winter) skimming the waters—all signs of the fishy riches lurking below the surface. When the scallop drags were dumped onto the decks, a rogue's gallery of sea-bottom creatures was revealed: yellow-and-orange-spotted deep-sea skates and rays, conger eels as thick as a grown man's arm, monkfish—"sea monsters" with enormous gaping mouths for swallowing prey whole—as well as various benthic invertebrates, including the sought-after scallops and a by-catch of lobsters and groundfish such as haddock.

The shallowness of Georges—in some places, such as Cultivator and Georges Shoals, it is only 3.65 meters (12 feet) deep—accounts in part for its productivity. Sunlight penetrates to the bottom of the bank, and tides and winds acting over the shallow banks stir the waters, causing vertical mixing that keeps the nutrients available. Furthermore, the clockwise gyre over the bank is a partially closed circuit, entraining the nutrients and plankton—and the fish eggs and larvae—for longer periods, thus boosting and sustaining productivity, estimated at four times that of the legendary Grand Banks off Newfoundland. Phytoplankton growth is fueled not only by nutrients supplied to the bank from the deep waters surrounding it but also by so-called regenerated nutrients, that is, those excreted from animals and bacteria. So the abundant life on the bank is, in part, a product of itself, or specifically, of its waste products, which are taken up and recycled in the system.

*Divvying Up the Riches*

Generally speaking, the Gulf of Maine is a meeting place for species native to areas north and south of it. On Georges Bank, for example, three biogeographic species groups mingle: Labrador species in the deep water, Acadian species at the intermediate depths, and Virginian species at the shallow depths on top of the bank. In summer, bluefin tuna pass through the gulf on spectacular migrations spanning the North Atlantic, and their smaller cousins, mackerel, leave the deep water off the Virginian capes to pass through the Gulf of Maine. Other southerly species, such as bluefish, butterfish, menhaden, and summer flounder, also come north to the bank in summer to feed and mix with northerly types such as the major groundfish species—cod, pollack, haddock, halibut, redfish, plaice, and argentine—which spawn on Georges Bank itself. Only two migrants from the north frequent gulf waters: Atlantic salmon returning from their Arctic feeding grounds to their natal rivers, and capelin, a subarctic member of the smelt family. Mixed in with these typically southern and northern species are more ubiquitous species that are regularly found in waters both to the north and south of the banks, namely, squid, lobster, scallop, herring, red hake, silver hake, swordfish, and yellowtail flounder, among others.

Fish are found in markedly different parts of the Gulf of Maine during different parts of the year. Pelagic (near-surface) fishes, like herrings, seem to be more common along the eastern parts of the coast and off New Brunswick, whereas demersal (bottom-dwelling) fishes, like cods and flounders, prefer the western parts of the coast. These distributions, it is believed, reflect the various mixing regimes within the gulf. In the west, the phytoplankton is produced in the spring and promptly sinks to the bottom, where it becomes fodder for demersal fish. But in the east at the mouth of the Bay of Fundy, where tidal mixing is stronger, the phytoplankton remains in the water column throughout the year and thus attracts the pelagic fish.

*following spread*: Fishermen dry their nets in hopes of a bountiful harvest.

The distribution of these major classes of fishes differs markedly also. Members of the cod family—cod, haddock, pollack, cusk, and hake—each have a preferred haunt. For example, haddock are more abundant at the eastern end of Georges Bank, whereas cod congregate in the nearshore waters of the western gulf. Similarly other cod-related species have their special places, separate from the others. In addition, different cod species spawn at different times of the year. This patchiness may be a mechanism that the ecosystem has evolved to protect itself from the vagaries of nature. It seems important to the integrity and sustainability of the ecosystem that there is a cod of some kind in abundance somewhere in the system at all times. This built-in resilience and adaptability of the system appears to have been undermined, however, by overfishing. In the early 1990s, groundfish stocks collapsed in the northwest Atlantic. In place of codlike fishes, dogfish and skates now hold sway, and whether groundfish stocks will recover is not known. If they do, it is likely to be a slow process.

DIFFERENT SPECIES OF whales also divvy up the banquet of marine resources in the gulf, according to their individual needs. Their travels and differing feeding strategies help to spatially integrate the ecosystem and give it greater stability. Together, the five large whale species that occur in the gulf probably number 3,500, far fewer than the 25,000 that likely existed before the colonial period of whaling. This difference translates as a 75,000-ton loss of whale biomass to the system. No doubt some of the energy that went into the growth of whales now fuels other marine life, but the absence of so much higher animal life must have had a significant impact on the ecosystem.

A humpback whale lunge feeds through a shoal of herring and krill.

It is no coincidence that all five species of great whales common to the gulf—the humpback, fin, minke, sei, and right whales—are baleen whales, adapted to feeding on the gulf's cornucopia of plankton. There are two general groups of cetaceans, an order of mammals that includes whales, porpoises, and dolphins: the odontocetes, or toothed whales, and the mysticetes, or baleen whales. "Mysticetes" derives from the Greek word for mustache, *mystax*, and refers to the hairy appearance of the whale bone, or baleen, that hangs from the whale's upper jaws. The baleen consists of a series of overlapping horny plates that are fringed on the inner margin with hairlike bristles. These bristles act as a sieve to strain food—zooplankton, shrimp, or small fish—from the water.

With the exception of the right whale, the whales in the gulf are all rorquals, a term derived from the Norwegian *ror*, tube, and *hval*, whale. *Ror* refers to the grooves or folds on the throat and chest. When the whale gulps water while feeding, these grooves expand enormously, like a giant accordion or bellows, and the whale then forces water out through the baleen by raising its tongue, retaining the food items in the baleen for swallowing. The right whale, however, has no throat grooves and is a more passive feeder, continuously straining out the food organism while swimming through swarms of small copepods or krill.

These whales are found in different parts of the gulf at different times of the year, depending on where the heaviest concentrations of food are. Today, the most endangered great whale on the planet is the North Atlantic right whale, with as few as three hundred to four hundred still in existence. Historically, they were hunted off Cape Cod and adjoining waters, and the few that remain can still be found in the spring in the Great South Channel region east of Cape Cod and Nantucket.

Satellite imagery demonstrates that the whales frequent very specific areas just to the north of a 100-meter (300-foot) depth line, along the northern border of Georges Bank. South of this line, strong tidal currents keep the water well mixed, and therefore colder, all year long. In spring, however, the surface water warms north of this tidal mixing front. Along this front, where warm and cold waters meet, zooplankton concentrates into extremely dense patches, and it is here that right whales can be found in mid- to late May. Whales also travel large distances within the gulf itself in order to satisfy their energetic requirements. They are thus able to exploit the patches of food on traditionally rich offshore fishing grounds as well as productive coastal areas, such as the mouth of the Bay of Fundy.

### *Rocky Bands*

The rocky shores that typify much of the coast of the Gulf of Maine north of Cape Elizabeth set it apart from the coastal regions to the south, which are rimmed by sandy beaches backed by lagoons and salt marshes. In northern New England, the glaciers scoured the shore of sediments, exposing the bedrock beneath. Daily, the receding tide exposes a vast area where plants and animals have adapted to living on hard surfaces that cannot be penetrated. Plants, such as the seaweeds, attach themselves to the substrate by holdfasts, and these gardens of seaweeds provide cover and nutrients for a large array of marine vertebrates and invertebrates, and make a significant contribution of energy to the coastal ecosystem.

The rocky shore, nevertheless, is an extremely stressful environment. Not only is it subject to the battering energy of waves, but the rise and fall of tide exposes the plants and animals, alternately, to daily wetting and drying. Furthermore, rainfall can cause dilution; sun, desiccation; and winter freezing and thawing cycles add further stress. Lacking root systems, the plants cannot adapt by allowing their exposed parts to die back. Neither can animals burrow to hide from these stressors as animals can in softer substrates like sand or mud. But, for the plants that attach themselves to the rocky shore and for the animals that are able to cling to it, there are certain advantages. Precisely because the environment is so stressful, there is little competition for the rocky real estate from other plants, and for the animals there are fewer species of predators. As well, the turbulent waters that ebb and flow over the rocky shores carry with them not only ample supplies of oxygen, carbon dioxide, and dissolved nutrients, but food for passive filter feeders such as barnacles and blue mussels. As a result, production along the rocky coast can be extremely high.

Along the rocky shorelines of the boreal coast, from Cape Cod north to Labrador, the intertidal habitat—the area bracketed by the highest and lowest tides—is characterized by distinct biotic zones where a typical community of plants and animals has adapted to the rise and fall of the tide. These organisms align themselves in more or less horizontal bands, which are examined below, in order, moving from the land toward the low-water mark.

The spray or splash zone, as the name implies, occurs above the tide line but is subject to marine influence when waves crash against the shore, sending their spray above the highest high tide mark. It is an ecotone—a transitional zone between the terrestrial and marine zones that belongs wholly to neither one. It occurs in areas exposed to the full brunt of the ocean swells, such as outer islands and rocky capes, rather than in more sheltered places, such as coves. In the latter case, terrestrial plants may grow to the limit of the highest spring tides. The plants that have adapted to the spray zone have a terrestrial origin but have evolved means to deal with salt. In particular, a number of species of yellow lichens flourish there, painting the often dull rocks with splashes of color. All lichen consist of two plant species, an alga and a fungus, functioning together in a symbiotic relationship. The fungus provides a moist protective environment for the alga by absorbing water, while the alga produces organic compounds through photosynthesis to nourish the fungus. The alga also has little holdfasts that glue the organism to the surface. Another so-called lichen that thrives in the spray zone is the bryozoan *Lichenopora verrucaria*, which grows in tarlike patches. Despite appearance, bryozoans are not plants but colonial animals. Bryozoa literally means "moss animals,"

A nor'easter spends its winter fury against the rocky shores of Acadia National Park in Maine.

and they are sometimes referred to as "sea mats." Individual colonies may cover several square centimeters, but it is difficult to know where one colony ends and another begins.

The black zone is the highest of the intertidal zones and is only covered by the tide twice daily during the highest course of spring tides. Nevertheless, unlike organisms in the spray zone, which are merely tolerant of salt, those that live in the black zone are dependent upon periodic exposure to the tides. This zone takes its name from the blue green algae that clothe the rocks in black, which Rachel Carson called "a dark inscription." The story they tell is of the evolution of organisms from the sea to the land.

Blue green algae are the most primitive of organisms. The most common species is *Calothrix*, a microscopic filamentous organism that, like bacteria, has no nuclear membrane. The cells produce mucilage sheaths that have a dual purpose, protecting the cells from drying out and, at the same time, glueing them to the rocks. These organisms cover the upper intertidal rocks in slippery mats, which are grazed on by the rough periwinkle.

Periwinkles are molluscs, a phylum that has two distinguishing anatomical structures—the mantle and the radula—found nowhere else in the animal kingdom. The mantle is a fold in the body wall whose function is to secrete the calcareous shell typical of molluscs. It also serves to house the gills and protect the larvae in their planktonic stage. The radula is a ribbonlike tongue, wound like a clock spring on the floor of the pharynx. It is made of chitin (the same substance lobster shells are made of) and is armed with sharp teeth arranged in several hundred rows. The radula acts as a conveyor belt for scraping algae from the rocks.

The highly vascularized lining of mantle cavity, which makes it akin to a lung, is the vital adaptation for living at the top of the intertidal zone, allowing the periwinkle to survive the period when it is not covered by the tide, which may be as long as a month. It can also breathe anaerobically—that is, without contact with the atmosphere—for up to a week when it seals its shell against the rock, a behavior sometimes necessary to prevent it from drying out. Unlike other periwinkles, it also produces live young. Eggs encased in a cocoon develop inside the mother and emerge as fully formed shelled creatures the size of coffee grounds.

In sharp contrast to the black zone, the barnacle zone is glaringly white because of the limestone shells of the barnacles. Rock or acorn barnacles are stationary and therefore must be exposed to the tide every day to feed. The barnacle is a crustacean; it glues itself to the rocks and waits for the tide to deliver its meal of microscopic life. It lives head-down in a volcano-shaped shell consisting of six calcareous plates. The top is fitted with a pair of trapdoors that shut at low tide to prevent the barnacle from drying out, but open as soon as water covers the shell. Then six pairs of feathery appendages called cirri appear and begin to sweep the water, creating a vortex that draws food particles into the crater where the little animal passes its sheltered life. The great 19th-century naturalist Louis Agassiz described the creature and its lifestyle well: "Nothing more than a little shrimplike animal, standing on its head in a limestone house and kicking food into its mouth."

Barnacles thrive in the upper intertidal zone and often monopolize it because of their many adaptive mechanisms for coping with its extreme environmental conditions. They can survive perhaps the greatest temperature range of any animal, from 44°C (111°F) to -15°C (5°F). At extremely low temperatures, barnacles and other invertebrates protect themselves against the

Barnacles monopolize the upper intertidal zone but for a starfish.

tissue-damaging effects of ice crystals by secreting antifreeze compounds, such as glycerol and trehalose, into their body fluids. In times of extreme heat, they cool themselves by means of transpiration, or evaporation of bodily water. When not covered with seawater, the barnacle can breathe humid air through a small passage, the micropyle, in its shell. And when the air is too dry, it can close its shell completely and survive on small amounts of stored oxygen.

The rocky coast at low tide reveals the bold banding typical of the Gulf of Maine and outer Bay of Fundy.

Its reproductive strategies further demonstrate the adaptability of this creature, which is one of the few crustaceans that is hermaphroditic—each individual possesses both male and female sexual organs. Each barnacle has a remarkably long penis, which can reach into the mantle cavity of a neighboring animal and fertilize it. The fertilized eggs are brooded in the mantle cavity until they are released as larvae and drift with the tide in search of a site to anchor themselves. Once they find a suitable site, they undergo a remarkable metamorphosis within a period of twelve hours: from a relatively shapeless larval form, the head and appendages emerge, and the cone of the shell is formed, replete with plates. During its three-to-five-year life cycle, the barnacle will have to molt its chitinous skin to grow, enlarging its shell to accommodate its growing body, leaving behind what Rachel Carson described in *The Edge of the Sea* as "semitransparent objects . . . like the discarded garments of some very small fairy creature." The molting process is usually triggered when the waters reach a particular temperature in the early spring, and all the barnacles in the region molt at about the same time.

Because of their remarkable hardiness, barnacles dominate the upper levels of the intertidal zone, where conditions are harshest. They do have predators, however, and, at the lower end of the intertidal zone, competitors.

AS WE MOVE farther down the rocky intertidal seashore, we enter the brown algae zone, which normally extends from the lower edge of the barnacle zone to the mean low water mark. The two most conspicuous plants along the rocky shore are types of brown algae, knotted wrack and bladder wrack, which cover the rocks like luxuriant heads of hair. Both plants have air-filled bladders along their fronds, which serve as flotation devices as the tide rises over their rocky homes. Bladder wrack is more dominant in energetic environments, exposed to wind and waves, and knotted wrack is better adapted to quieter, more sheltered places. These plants—the fucoids—have also evolved strategies to cope with the stresses that come with a territory where temperatures fluctuate and desiccation is a threat. Chemicals in their cell walls, called alginates, counter these stresses but also lend flexibility and strength to the plants as a buffer against the waves' often violent force.

Barnacles can live at all levels of the intertidal zone, but in the brown algal zone, new organisms outcompete the barnacles, which are so dominant in their own domain. The algae overgrow the barnacles and impede their ability to feed, thus reducing their numbers. The rock weeds also shelter the primary predator of barnacles, the dog whelk. And the blue mussel outcompetes the barnacle, largely replacing it in the brown algae zone.

Blue mussels colonize an area faster than any other organism, in part because of their great fecundity. A female produces up to 12 million eggs, which are fertilized when they are broadcast widely into the water column. The larval mussel settles to the bottom, where it metamorphoses into a miniature mussel and moves up the shore looking for a place to attach itself. It does so by means of byssal threads produced from a protein substance in its foot. These act as cables or guy wires, anchoring it to the rocks. Like most bivalves, mussels are filter feeders, drawing in water through one siphon and passing it over mucus-coated gills that trap the food particles as well as extract oxygen, and then expelling the cleaned water through a second siphon.

Populations of the fecund blue mussels and barnacles are held in check by the predatory dog whelk, whose beautiful spiral shells are a collector's delight but a mussel's or barnacle's nightmare. The dog whelk pries open the top trapdoors of the barnacle, then inserts its radula between them. It does not accomplish this feat by sheer physical force, however. Covering the barnacle with its foot, it first secretes a highly poisonous purple dye, purpurin, once prized by native North Americans, to kill the barnacle. It adopts a different strategy with the much larger and stronger mussel. It drills a small hole in the shell, then inserts its proboscis, which houses its mouth. Again, however, it gains entry by first secreting a substance that softens the shell, then uses its many toothed radula to drill a larger hole.

These dramatic predatory acts take place under cover of the fucoids. The luxuriant growth of these seaweeds has evoked the terrestrial metaphors of a "forest" or even a "jungle." The comparison is also apt because of the diversity of animals that live there.

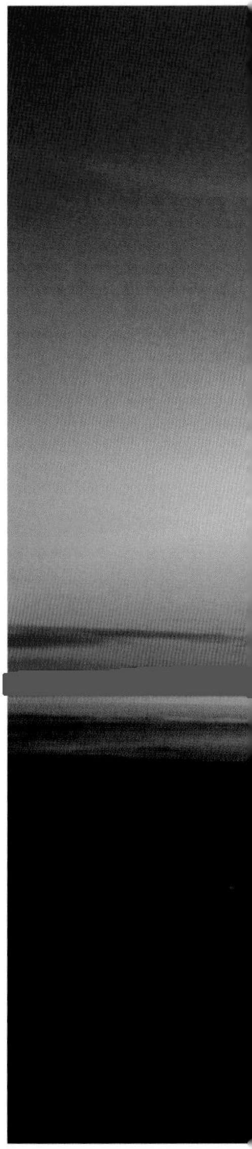

These plant communities perform a number of vital ecological functions. At high tide, they act as canopies that shade and shelter some twenty-two juvenile fish species, including tomcod, pollack, sculpin, cod, alewife, white hake, and flounder. At low tide, the canopies become wet mats that protect many species of tiny invertebrates, which, in turn, attract some fifteen species of seabirds and shorebirds in search of food. The most common creature sheltering under the fucoid forest is the amphipod *Gammarus oceanicus*, or as it is more commonly known, "sideswimmer" or "scud." Pull apart any clump of rockweed, and hundreds of these crustaceans scatter for cover. These rockweeds also contribute to offshore productivity when parts of the plants break off and drift seaward. There, they can form extensive mats that shelter zooplankton, which are eaten by larval lobsters and juvenile fishes such as lumpfish, sticklebacks, rockling, and hake, as well as by seagoing birds like phalaropes and terns.

Below the brown algae, in the low intertidal zone, we enter the red algae or Irish moss zone. The namesake plant is Irish moss, a ruddy, many-fingered plant that forms dense mats, especially on flat rocks, where its tenacious holdfasts allow it to flourish in this turbulent zone. (Processed Irish moss is a common ingredient in ice cream and other foodstuffs. Another edible red algae is dulse, which is sun-dried and eaten as a snack or condiment.) Blue mussels may occupy areas where the Irish moss fails to take hold. The

The erosive power of Fundy's famed tides have carved these curious "flower pots" from the inner bay's soft sandstones.

common periwinkle is a frequent grazer in the red algae zone but concentrates on ephemeral species, to the benefit of the Irish moss. The green sea urchin feasts on Irish moss, however, as well as devouring periwinkles and even mussels and barnacles.

*Seabird Republics and Seal Rookeries*

Some five thousand islands lie in the Gulf of Maine off New England's rocky coast, but only 10 percent of them provide potential nesting sites for gulls, terns, cormorants, petrels, guillemots, and other seabirds. Seabirds require

Dapper razorbill auks stand guard over Machias Seal Island, the most southerly breeding colony for the species in the world.

three conditions to raise their young safely. First, the island must be free of large mammals, like foxes, mink, coyotes, and rats, which can prey on their eggs, their hatchlings, and the adults themselves. Second, most seabirds prefer an open habitat, free of trees, for nesting. Finally, the islands must be close to abundant offshore resources. These conditions are met mostly on the outermost islands and rocks around the gulf, including Machias Seal Island at the mouth of the Bay of Fundy; Schoodic and Great Duck, off Mount Desert Island; Matinicus Seal, Matinicus Rock, and Metinic Islands in outer Penobscot Bay; and the sandy islands off Cape Cod, such as Manomet and Monomoy. The most important of these islands is Machias Seal, which may be the most important seabird-nesting colony south of Newfoundland.

I visited Machias Seal twice in the 1980s, and on both occasions I heard the island before I could see it through the thick shroud of fog—a product of the bay's cold waters and warmer summer air—that often drapes this 10-hectare (25-acre) hunk of wave-washed granite. The murmurings and cries of the seabirds mingled with the shushing sound of the surf. On the two-hour trip from Grand Manan Island, I spied Atlantic puffins on the water with several sardines draped in their colorful, parrotlike beaks, and looked up to see the graceful silhouettes of terns as they hovered on their delicate, backswept wings. Once the island came into view, groups of razorbill auks, relatives of the extinct great auk, aligned onshore in a seemingly formal pose in their black-and-white plumage, along the granite boulders that form the island's natural breakwater.

Traditionally, some 2,800 Atlantic puffins have nested here, making it the largest colony of puffins south of Great Island, Newfoundland. It is also the most southerly breeding colony in the world of razorbill auks, with 900 pairs joining their alcid cousins, and historically, it has been host to the largest colony of Arctic terns in eastern North America. In the past, approximately 2,800 pairs of Arctic terns were joined by 1,330 pairs of common terns, as well as small numbers of Leach's storm petrels. In recent years, however, the number of terns nesting on the islands has dropped drastically, and since 2006 terns have failed to nest on the island at all. Changes have also been observed in the diet of the alcids, with sand lance, krill, and fish larvae replacing juvenile herring as the most important food item. Studies on the island continue in an attempt to explain these trends, which may be related to a number of factors, including overfishing of herring stocks, increased predation by gulls, and climate change.

IN ADDITION TO the many islands rimming the coast of the gulf, many more rocky ledges may be inundated by the tide. These are often haunts for harbor seals, which can often be seen basking there in groups. Although they have been persecuted by humans because they compete for fish and carry a parasite that infected cod, harbor seals show a remarkable curiosity toward humans, often hauling out onto beaches for a better look at us. In ancient times, this behavior gave rise to the selkie legend, which purported that seals could assume human form, and vice-versa. Like seabirds, seals must come ashore, or onto the ice as northern species do, to give birth. They also haul out onto land in order to dry out and to rest, often in company with other seals. This gregariousness and the need to breed on land have led to the evolution of breeding colonies.

The many rocky ledges that rim the gulf shore provide ideal breeding and whelping places, or rookeries, for the thirty thousand harbor seals native to the region. From spring through autumn the New England population is spread along the Maine and New Hampshire coast in several hundred colonies, the largest containing 150 or more individuals. The greatest concentration of

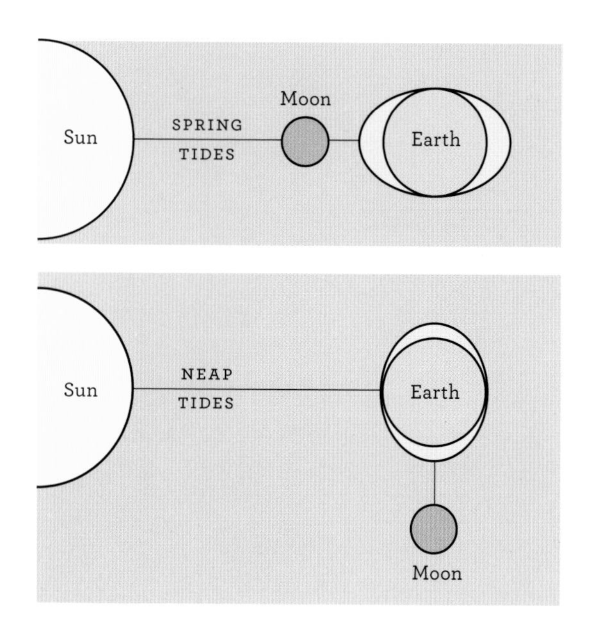


When the sun and the moon are aligned, they produce the highest, or spring, tides; when they are at right angles to each other, the result is the neap, or lowest, tides of the month.

seals is in Machias and Penobscot Bays, and off Mount Desert and Swan's Island in the Acadia National Park area. Males and females breed promiscuously in September and October.

Although some harbor seals overwinter in the gulf, by late autumn most migrate southward to winter off the shores of the Mid-Atlantic from Cape Cod to Chesapeake Bay. They return in spring to give birth in May and early June in the protected waters of the gulf's many embayments, showing fidelity to particular ledges. Adults molt in July and August, acquiring a new coat that appears pale and silver when dry, after which some move to ledges farther offshore, where there is more food in the deeper water. Their most common prey are herring, alewife, flounder, and hake, but they are opportunistic feeders that will also prey upon squid and other invertebrates.

*Bay of Giant Tides*

Where the Gulf of Maine ends and the Bay of Fundy begins is difficult to say, though Canada and the United States draw a national marine boundary near Machias Seal Island, each nation claiming this seabird republic for its own. But there can be no doubt that the best place to appreciate the power and volume of the Fundy tides is Cape Split, an imperious 100-meter-high (300 feet) headland overlooking the entrance to the Minas Basin in the inner Bay of Fundy. Here, a volume of water equal to the Gulf Stream, or two thousand times the discharge of the St. Lawrence River, must squeeze through the 5-kilometer-wide (3-mile) Minas Channel. It does so under furious protest, creating huge standing waves, eddies, and whirlpools that resound and reverberate at the base of the cliffs as the tide floods into Minas Basin, where it can reach a height of 16 meters (50 feet) or more—the highest in the world.

As most of us have been told, the moon's gravity causes the tides, but not simply by lifting the waters of the Earth. The Earth responds to the gravitational pull of the moon, in effect, by falling toward its celestial neighbor. The waters on the side nearest the moon, being closer, accelerate more quickly than the Earth itself. For the same reason, the Earth falls more quickly toward the moon than the waters on the far side, so that the Earth is being pulled away from these waters. The result is that two tidal bulges are produced on

## MAINE'S MANY ISLANDS

MAINE BOASTS more islands than anywhere else on the Atlantic coast, with 4,617 islands listed in the state's Coastal Registry. Their granitic foundations were formed by molten rock that bubbled up from deep within the Earth when the European and North American Plates collided 430 million years ago. Today, the islands themselves are formed of drowned ridges, hills, and mountaintops, but most have a low profile, since this coast has rebounded only relatively recently from glacial pressure. An exception is Mount Desert Island, whose pink granite heights at Great Head rise 44 meters (145 feet) above the pounding surf. Outer islands are exposed to the full fury of the infamous nor'easters. On Monhegan Island—the name derived from the Algonquian, "out-to-sea-island"—such storms send a veil of spray over top of the 30-meter-high (100-foot), spruce-clad White Head.

Monhegan was an important landfall for early sailors and fishermen. In 1605, the explorer George Weymouth described it as "woody, growen with Firre, Oke, and Beach, as far as we saw along the shore . . . The water issued forth down the Rocky Cliffs in many places; and much fowle of divers kinds breed upon the shore and rocks." Still today, the island shelters a hardy community of winter fishers.

The many islands along Maine's rocky coast "look like so many wood chips scattered across the water," the dean of New England natural historians, John Hay, once wrote. This dense scattering of islands has an important biological effect, boosting productivity along the entire coastline.

As the water of the Gulf of Maine circulates among the islands, they help to mix and oxygenate it, as well as causing local upwellings that draw the colder, nutrient-rich water from the bottom into the photic zone. The volume of food available to filter feeders and their predators is also increased as tide-driven currents are funneled between the islands.

Islands significantly increase the overall length of the Gulf of Maine coast, much of which is clothed in seaweeds that serve as the principal primary producers in the inshore environment. Lobsters, crabs, and fishes make inshore migrations in spring and summer and find shelter in the underwater rockweeds and kelp forests surrounding many of the islands. This habitat is particularly important in the life cycle of lobsters, when they abandon their deep water winter grounds and begin to shed their shells. During this vulnerable period, rockweeds provide cover and a rich source of food as the lobsters transform themselves from scavengers to filter feeders.

For these reasons, Maine islands are frequently rimmed by a garland of lobster traps, and in fact the largest lobster harvests are found along sections of the coast with the largest number of islands.

opposite sides of the Earth. As the Earth rotates under these bulges, it produces the twice-daily, or diurnal, tides typical of the world's oceans.

As coast dwellers know, tides vary in height during different times of the month and year, and this variation also relates largely to the Earth-moon celestial relationship, with the added influence of the Sun. At 40 percent that of the moon, the sun's effect on tides is relatively large, given that it is so much farther away. Every two weeks, the sun, moon, and Earth are aligned, maximizing their combined effect. The result is the highest tides of the month—the spring tides. (The term does not apply to the seasons but derives from the Anglo-Saxon word "springen"—to leap up.) Conversely, at the first and last quarter of the moon, the solar and lunar influences are at right angles to each other, diminishing their combined effect and producing the smallest tides of the month, or neap tides. The extremes of tides, both high and low, vary over the course of the year, again as a result of the relative positions of, and distances between, sun, moon, and Earth.

*facing page:*
Maine's many islands–some 4,600 of them–boost productivity in the gulf by creating local upwelling of nutrients from the sea bottom.

These astronomical factors influence the range of tides the world over. It follows, then, that the conditions that produce the extreme tides in the Bay of Fundy have a local origin. The laws of physics tell us that every basin has a characteristic period of oscillation, which means that once set in motion, waters within it will slosh back and forth with a regular rhythm. Where an open bay communicates with the sea, however, the influence of the lunar-dominated tides must also be taken into account. If the basin and ocean tides are rocking back and forth in harmony with one another, they are said to be in resonance. This is what happens in the Bay of Fundy and largely accounts for the great tides experienced there.

When applied to water in motion, resonance is more commonly referred to as the "bathtub effect." A home experiment can easily demonstrate what happens on a grand scale in the Bay of Fundy. Draw a shallow bath and give a push to the water at one end of the tub. A wave will travel to the other end and be reflected back. Just before it reaches your hand, give another push and you will observe that the water begins to slosh back and forth quite dramatically. In the bay, the moon acts as the hand in the bathtub analogy; it has been shown, however, that it is not the bay alone but the Bay of Fundy and the Gulf of Maine acting as a single oceanographic system that produces this effect. The period of time it takes for the tidal wave, or bulge, to travel from the mouth of the bay to the head is only nine hours—not long enough to cause resonance. When the bay and the gulf are considered together, the resonant period is 13.3 hours, close enough to be in resonance with the 12.42-hour, moon-forced tidal system.

The tidal range increases progressively from the mouth of the bay between Digby, Nova Scotia, and Saint John, New Brunswick, where it reaches heights of 3.5 meters (11.5 feet), to the head of the bay, where so-called mega-tides, in the range of 14 to 16 meters (46 to 52 feet), occur. As the tide flows into the bay—a volume of water equal to all the rivers on the planet combined—it is forced into a shallower and narrowing space, and, in effect, piles up. This pile-up produces in some of the tidal rivers at the head of the bay a hydrographic phenomenon known as a tidal bore. A bore is the leading edge of the advancing tide that forms a tumbling wave front, more than a meter (3 feet) high. It often travels far inland before dissipating when it is finally overwhelmed and dampened by the volume of rising water behind it.

*Upwellings of Life*

The Bay of Fundy has been aptly described by biologist and bay expert Graham Daborn as "an ecosystem with a biological pump at both ends." In the lower end, or outer bay, the tide rips—which gave their native son Joshua Slocum such a thrashing as they boiled over the reefs around Brier Island—pump nutrients from the seafloor up into the photic zone. In the upper end, or inner bay—the part of the bay that includes the Minas Basin and Chignecto Bay—salt marshes and mudflats act as the biological factories, pumping their production into the marine zone as the tides ebb and flow.

The turbulence that characterizes the waters at the mouth of the Bay is associated with an oceanographic phenomenon known as upwelling. Upwelling usually occurs near the coast, where prevailing winds move waters away from shore, and colder, deeper waters move upward to replace them, bringing their bounty of nutrients. This phenomenon is most frequently observed on the west coasts of continental margins, notably off Oregon and Washington, Peru and Chile, and Morocco and southwest Africa. Most upwellings are wind-generated. In the Bay of Fundy, however, the prevailing winds shift with the seasons, from northwesterly in winter to southwesterly in summer, suggesting that upwelling is not primarily the result of wind action. In Fundy, as should be no surprise, it is the bay's great tides themselves that are the generators of upwelling. As the strong tidal currents pass around the corner of the coastline near Yarmouth, at the southwestern tip of Nova Scotia—which is oriented at right angles to the direction of the current—they create an upwelling effect in the same way that turbulence is created when a river flows around a bend. Further, as the tide streams around the bend of southwestern Nova Scotia, it runs up against horseshoe-shaped ledges that produce the frightful rips near Brier Island—which are, in fact, miniature upwellings.

A herring weir is strategically located to catch juvenile herring along the New Brunswick shore, where they are canned as sardines.

The visible result of these upwellings is not only turbulence and tide streaks but also a remarkable concentration of marine life, from the smallest to the largest creatures found in the sea. Most notably, whales and seabirds come to take advantage of the food-concentrating power of the tides. The copepod *Calanus finmarchicus* is a big part of the attraction. This weak swimmer is brought to the surface by upwellings, where both red and red-necked phalaropes—sea-going shorebirds—feast upon it. Until the 1990s, as many as a million red-necked phalaropes congregated in the Passamaquoddy Bay area, while on the eastern side of Fundy, near Brier Island, some ten thousand red phalaropes returned each year to feed during their north-south migrations in late summer. These great flocks have largely disappeared in the last two decades for reasons that are not as yet clear but may be related to nodal cycles, in which tidal amplitudes tend to increase for 9.3 years, then decrease during the next 9.3 years. This 18.6-year-cycle relates to the variation of the moon's orbit about the Earth relative to the Earth's about the sun.

## THE COMPLEX LIFE OF A LOBSTER

CONSIDERED BY most as a culinary delicacy and luxury, lobsters are the most sought-after seafood along the eastern seaboard. This was not always so, however. In the 18th century, in Massachusetts, it was forbidden to serve lobster to prisoners and servants more than twice a week, and until the mid-20th century, poor schoolchildren in Maritime Canada hid the fact that they had to eat lobster sandwiches for lunch. The American lobster is found along the coast between Cape Hatteras and the Strait of Belle Isle between Newfoundland and Labrador but is especially abundant in the Gulf of Maine, along the Atlantic coast of Nova Scotia, and in the southern Gulf of St. Lawrence.

As the bearers of a flexible shell, lobsters belong to the class Crustacea and are cousins to crabs, shrimp, and copepods. They have five pairs of legs, including an exaggerated pair of asymmetrical claws at the front of the body: the crusher and the pincer, used for defense and food gathering. The smaller legs are covered with hairs, which function as taste organs, and the compound eyes, consisting of 13,500 light-capturing, image-forming organs called ommatadia, are mounted on flexible stalks at the front of the body. It is not known whether nocturnal lobsters perceive, or need to perceive, color, but they have a highly developed sense of smell, with the first antennae acting as the "nose."

Lobsters are bottom-dwelling creatures that generally live in water depths of less than 50 meters (164 feet) and prefer cobble habitat, where as juveniles they seek shelter from predators in the crevasses among algae-covered stones. As adults, lobsters have few enemies other than humans, and it is believed they can live fifty years or more. But first they must survive a series of vulnerable stages in their complex life cycle.

Shortly before she must shed her shell, the female initiates the mating ritual by poking the tips of her claws into the shelter of the male, with whom she will form a brief pair bond of a week or two. After a few days of this tentative shelter-checking, she enters the male's hideout, where he taps her claws in a seeming welcoming gesture. This "boxing" behavior probably has the more practical function for the male of determining how soon the female will shed her shell. When she does, she turns over so that her abdomen faces the male, who transfers the sperm packet into her using his first pair of swimmerets. The female will stay with the male for a few days after mating, during which the male provides protection while her shell begins to harden. She then seeks a shelter of her own until her shell hardens sufficiently, which may take months, at which time she lays her eggs. In most cases, the eggs are not laid until a year after mating, and the female then broods them on her swimmerets (or pleopods) for nine to twelve months.

Lobsters are especially abundant in the Gulf of Maine and at the mouth of the Bay of Fundy, where they support a lucrative fishery.

When the eggs are ready to hatch, she releases the larvae over a period of several nights to several weeks, likely to minimize the risk of losing all of the young to predators at once. These larvae float to the top of the water and become part of the plankton. Resembling small shrimp, they undergo three molts, and during the last of these they metamorphose into a form resembling tiny lobsters (the so-called postlarval stage). They then settle to the sea bottom, where they will spend the rest of their lives. There they remain hidden, assuming a cryptic lifestyle, until they reach a size between 25 and 40 millimeters (1 to 1.5 inches), when they become more mobile (a stage dubbed "vagile"). Out of the ten thousand larvae released into the water column, only one will survive to adulthood—and then might reach some connoisseur's plate.

Bonaparte's gulls are among the seabirds that benefit from the "tidal pump" in the outer Bay of Fundy.

Other seabirds, however, still converge on the rich waters of the outer bay. As many as ten thousand shearwaters concentrate around Brier Island, and on the New Brunswick side of the bay, Bonaparte's gulls use Passamaquoddy Bay as a major molting and staging ground during fall migration. Later, the Grand Manan Basin becomes a major wintering area for razorbills and their auk cousins, common murres. Cetaceans, too, seem to divide up the territory. The North Atlantic right whale and harbor porpoises concentrate in the Grand Manan Basin, and humpback whales are more common on the Nova Scotia side. This separation of species is probably some expression of feeding ecology related to the size of the zooplankton prey. In seabirds, it may be related to beak size, whereas in whales the fineness of the baleen fringes may determine what species are found where.

Fish also concentrate in these areas where feed is pumped to the surface. Two-to three-year-old juvenile herring, in particular, feed along the New Brunswick shore and near Digby Neck on the Nova Scotian side of the bay, where they support a sardine industry based on small herring. I have watched as Atlantic herring were pursed from a fish weir off Grand Manan Island, at the mouth of the Bay of Fundy. A net is strung around the inside perimeter of the weir, itself a giant wood and twine basket with an ingeniously designed "door" allowing the unwary herring to enter it but not to exit. The net is then closed, or pursed, by a draw string at the bottom, and raised, or as local fishers say, "dried up." It is a descriptively apt phrase, for as the net is pulled up, the water itself is seemingly transformed into a seething, electrifying mass of herring. Not only does this silvery bounty support many coastal communities along the Bay of Fundy and Gulf of Maine, but herring, in its many life stages from egg to adult, is a cornerstone species for the entire marine ecosystem.

Herring consume plankton and in turn are consumed by the higher predators. Their sheer numbers—some 40 billion individuals in schools covering several square kilometers, so dense they block out the sun for divers below—underscore their importance. Harbor porpoises' predilection for herring has earned them the sobriquet "herring porpoises," and the availability of young herring for breeding seabirds, such as puffins, razorbills, and terns, is critical to the survival of their young on the breeding islands in the Gulf of Maine-Bay of Fundy system.

Humans corral these massive shoals in weirs and seines, but humpback whales have also learned to trap herring by blowing bubble nets. I have stood on a boat deck near Brier Island and looked overside as a group of humpbacks blew a net of bubbles around a school of herring, then—lunging from below, mouths open—exploded to the surface to engorge and swallow hundreds of the silvery morsels at once.

MOST NOTABLY, THE ecologically unique approaches to the Bay of Fundy are a magnet for the North Atlantic right whale, the rarest of the world's large cetaceans. *Eubalaena glacialis* got its unfortunate epithet as the "right whale" to hunt because it produced large quantities of oil and conveniently floated to the surface when killed, allowing for easy retrieval. We know that the Basques began to hunt them in the eastern North Atlantic in the 11th century and moved their whaling operations to the New World in the 15th and 16th centuries. They established a series of whaling stations along the Labrador coast to intercept the bowhead and right whales migrating through the Strait of Belle Isle, where they harvested as many as twenty thousand of these two species. In the era of industrial whaling, the right whales were pursued relentlessly by New England whalers for their highly prized oil, used initially as lamp fuel and later as industrial lubricants, and their flexible baleen plates were a precursor of plastics, employed for everything from corset stays to buggy whips. Their populations were so decimated by the dawn of the 20th century that they became "commercially extinct," and therefore unprofitable to pursue. Belatedly, in 1935, the League of Nations declared it illegal to hunt them.

*facing page:* The Bay of Fundy and Gulf of Maine are critical feeding grounds for the rare North Atlantic right whale (top) as well as home to the more common harbor seal (bottom).

Although other protected whale populations, such as the Pacific gray whale, have made dramatic recoveries, the North Atlantic right whale remains critically endangered. Records of its seasonal occurrence off Cape Cod date to the mid-16th century. In the late 1970s, scientists working with ships' officers on ferries, in a landmark example of "citizen science," observed fifteen right whales in the bay—a revelation greeted with skepticism at the time. But in the 1980s scientists determined that groups regularly returned to the outer Bay of Fundy, in Passamaquoddy Bay and around Grand Manan Island. Research has since established that these areas in Fundy, as well as Roseway Bank on the Scotian Shelf, are critical to the survival of the species. That survival is precarious, as the population has been reduced to roughly 350 animals, making it the rarest of the rare.

Every summer, the females bring their calves to these food-enriched waters. It now appears that the coastal area of the southeastern United States, between Georgia and Florida, is a major calving grounds, but these southern waters are relatively sterile compared with the colder waters to the north. In spring, after birthing, the right whales undertake a 2,900-kilometer (1,800-mile) seasonal migration northward, during which they stop off at traditional feeding areas along the coast, such as the Great South Channel between Cape Cod and Georges Bank. By mid- to late June, the whales begin showing up in the outer Bay of Fundy and on the southern Scotian Shelf, where they spend the summer laying on fat.

This seemingly steady movement is marked by an underlying urgency. The females have not fed while on the southern calving grounds, and they now need to replenish their own bodies and provide nourishment for their newborns for the first nine months of life. To do so, they need to exploit areas with high densities of copepods, such as the Bay of Fundy. Fundy is a dual-purpose area, serving as a nursing or training ground for young calves and as an important feeding ground for calves and mothers. Normally, right whales are skim feeders: they simply plow along the surface through marine meadows of copepods, straining the tiny zooplankton through their baleen plates. In the Grand Manan Basin, however, the whales employ a novel strategy, making deep dives to exploit the copepod masses that are concentrated near

the bottom by a combination of ocean fronts and a local gyre that roughly follows the figure-eight shape of the Grand Manan Basin. Downwelling occurs at the edge of the basin, where well-mixed and less well-mixed water masses converge. At this transition zone, the copepods settle out—in effect, they slide down the transition zone as down a pane of glass—yielding a rich vein of the energy-rich zooplankton. Right whales' exploitation of the region is greatest in August and September, to coincide with the peak pulses of copepod production.

The Bay of Fundy may always have been a natural refuge for the species. Historically, the frequent fogs and the treacherous tidal currents discouraged whaling in the bay during the industry's heyday. Certainly, the bay will play a vital role in the right whale's precarious future. Cows with calves show a marked preference for the bay over other areas, such as the Great South Channel and Scotian Shelf, even though they also offer a rich copepod diet. There may be two other reasons for this: Fundy is well protected from storms, and it enjoys a relative lack of predators such as the killer whale, both threats to vulnerable calves.

### *Marshes, Mudflats, and Migrants*

The biological factories, or pumps, in the inner Bay of Fundy are the lush ribbons of salt marsh, nourished by the tides, and the vast mudflats that are exposed when the giant tides ebb away. In sharp contrast to the blue waters of the outer bay, those in the inner basins (Chignecto and Minas) are a muddy red. The turbulence of the great tides scours the sea bottom and so muddies the waters that the biological production by phytoplankton in the water column is limited. Nevertheless, it is an illusion that the reddish brown waters of the inner bay are barren. Over the centuries, these same tides have built up large salt marshes, which, in turn, export their organic production into the marine zone. The mudflats exposed by the ebbing tide are natural solar panels covered in slicks of diatoms—single-celled algae, each acting like a solar greenhouse to absorb the sun's energy and convert it into organic compounds. Tides churn this production into a nutrient soup, which feeds a variety of bottom-dwelling creatures, fish, and birds.

The most important nutrient "pump" in the inner Bay of Fundy is the salt marsh, which is the source of half of the primary productivity. Salt marshes in the bay and elsewhere in the Gulf of Maine are divided into low marsh and high marsh. Low marsh is flooded twice daily and is dominated by the cordgrass *Spartina alterniflora*, whereas the high marsh may be flooded only a few

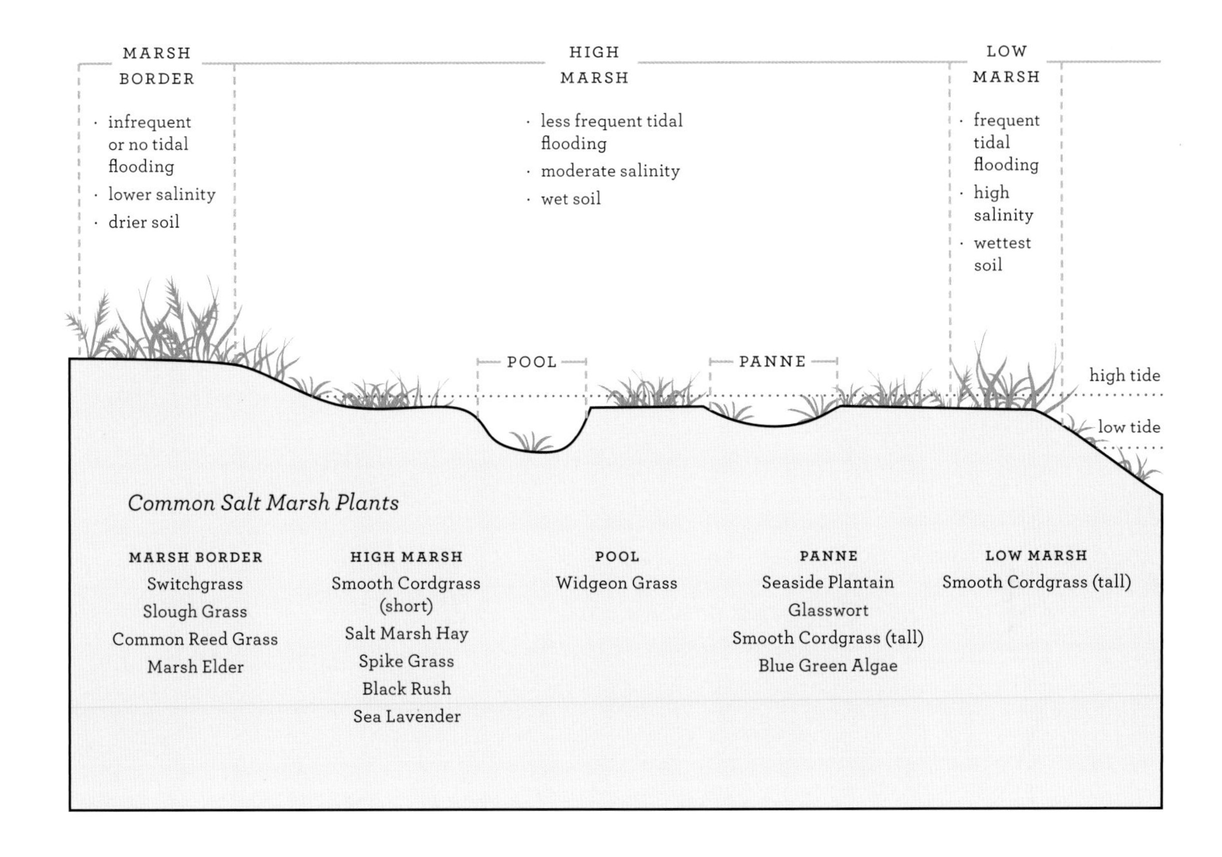


Zonation of salt marsh plants–on the low and high marsh–depends upon the frequency of tidal flooding and the plants' varying tolerances to salt.

times every month during the spring tides and is dominated by a shorter, less luxuriant species, *Spartina patens*, or as it better known, salt marsh hay. Both of these cordgrasses have one thing in common—an ability to cope with salt, which effectively excludes most other land plants from growing in areas flooded by the tide. When most plants are exposed to excess salt, water flows out of their cells in an attempt to equalize osmotic pressure on either side of the cell membrane. As a result, the cells wither and die. Salt marsh cordgrasses cope with the increased salt concentration of seawater by actually drawing salt water into their cells and concentrating the salts there in an attempt to equalize the osmotic pressure. If they retained too much salt, however, water would flow into the cell uncontrollably, with deadly consequences. To rid themselves of excess salt, the cordgrasses are furnished with specialized glands in their leaves, which excrete the salts.

In the low marsh, *S. alterniflora* creates a natural monoculture, though occasional glassworts grow on bare sites created by ice scouring, for instance. On

The extensive mudflats and salt marshes of the inner Bay of Fundy combine to boost biological productivity.

the high marsh, the salt marsh hay is joined by a number of other salt-tolerant plants, such as salt marsh bulrush, marsh elder, sea lavender, seaside aster, and seaside goldenrod.

Most of this plant matter is shed by the action of the tides, or by ice in winter, into the marine zone, where it becomes a pool of nutrients—feeding both the residents of the system and the migrants into the bay. Once, 300 square kilometers (115 square miles) of salt marsh fringed the Bay of Fundy coastline, but beginning in the 17th century the French Acadian settlers began dyking the marshes to create agricultural crop land, a practice that continued into the 20th century. Today only 15 percent of the original salt marsh is still open to the sea. Even with this significant loss of potential productivity, the inner Bay of Fundy remains a remarkably vital ecosystem.

The inner bay not only nourishes native species but also hosts remarkable immigrations and is a critical feeding ground for at least two populations: the American shad and the semipalmated sandpiper. Like whales in the outer bay,

ATLANTIC STURGEON

these species come to the inner bay to feed. Lying at the center of their migrations, Fundy functions as a critical link in their life cycles, tying together the Northern and Southern Hemispheres.

The American shad spawn in rivers all along the Atlantic coast. The translucent larvae spend their first summer in the river, feeding on planktonic crustaceans, and in the fall, when the temperatures drop below 15°C (59°F), the 10-centimeter (4-inch) fry descend to the sea.

Just where in the sea they go was a long-standing mystery in fishery science. Studies of Connecticut River shad first showed that they made impressive coastal migrations, moving northward as summer advanced along the coast. Shad with origins as far south as Florida were found in the Gulf of Maine. These migrations take place in response to changing water temperature, with shad seeming to prefer to stay in an envelope of water (known as their isotherm) with a temperature range of 13 to 18 degrees. This predilection for warm water leads them into the Gulf of Maine and ultimately into the Bay of Fundy, the only places along the Atlantic coast with that specific temperature range during August and September.

During the five years that shad normally spend at sea, they may travel 20,000 kilometers (12,400 miles), appearing as far north as Nain, Labrador, in summer, and as far south as Florida during January and February. All shad from the eastern seaboard enter the Bay of Fundy at least once during their life cycle, and tagging experiments in the inner bay have shown returns from every river from Florida to Labrador that contains a known spawning population of shad.

It is estimated that there are 10 million adult shad and many more juveniles in the Bay of Fundy every summer—half the North Atlantic population. While in the bay, they feed on mysid shrimp and other small crustaceans that, in turn, are nourished by the detritus-based food system originating in the salt marshes.

In addition to shad, a parade of southern fishes, including striped bass, alewives, sturgeon, and dogfish, come up the eastern seaboard and into Fundy's tidally dominated migratory circuit to exploit this summer feeding ground.

As well, in season, the tides become thoroughfares of indigenous fish life, as anadromous and estuarine species, such as tomcod and smelt, make their spawning runs into the tidal rivers at the head of the bay. Other deep-water species—cod, pollack, mackerel, and halibut—also visit Fundy's shallow upper reaches during certain months to feed on the offerings of mudflat and salt marsh.

IN JULY AND August, the inner basins of the bay also play host to another remarkable influx of life—feathered migrants from the north. Some 2 million shorebirds converge at a select number of sites to feed and fatten during their journey from their Arctic breeding grounds to their wintering grounds in South America. Of the thirty-four species documented in the inner bay, by far the most numerous is the semipalmated sandpiper. Up to 95 percent of the world population of these sparrow-sized "peeps" depends on Fundy's inner basins for the energy required to make the epic migration.

There is no more awe-inspiring spectacle in Fundy than the flight of sandpipers over the vast mudflats, sometimes in massive flocks of a half million. These brown and white birds perform an aerial ballet as they bank and turn in unison, flashing white and dark like two sides of a mirror. It is a living light show, as the birds fly wing-to-wing in an impossibly intricate choreography.

Most of their time in Fundy is not spent in flight, however, but on the mudflats, where they follow the advance and retreat of the tide line, obsessively feeding with their characteristic sewing-machine motion. Their prime prey is the scavenging amphipod *Corophium volutator*. Commonly called the mud shrimp, *Corophium* is a tiny, lipid-rich crustacean that feeds on the benthic algae which overspread the mudflats like a living membrane. Common to the European coast, it is found on this continent only in the Bay of Fundy–Gulf of Maine region. (The organism may have been an early invasive species that arrived in ballast during the Age of Discovery. That speculation raises the question, however, of how a native shorebird could have evolved such a dependency on an invasive species in just a few hundred years.) In Fundy, the number of mud shrimp in a particular mudflat is directly related to the amount of very fine sand. *Corophium* builds its U-shaped burrow with seemingly biblical wisdom: too much fine sand and its burrow collapses, too little and the mud is too thick to dig. Where substrate conditions are ideal, the density of *Corophium* peaks at an astronomical 63,000 per square meter and averages an impressive 20,000. Shorebirds forage in greater numbers on mudflats with the highest densities of *Corophium* and avoid mudflats where it is scarce, even though other prey, such as saltwater worms, may be extremely numerous—proof that they are critically dependent on the mud shrimp as their prime source of caloric energy for their impending flight to South America.

The key to the successful functioning of this prey-predator relationship is the reproductive cycle of *Corophium*, which undergoes a population explosion just in time for the arrival of the sandpipers from their Arctic breeding grounds. In Europe, *Corophium* produces only one generation of young, whereas in Fundy, two generations are produced annually. *Corophium* live but a year, and in Fundy, ice-scouring drastically reduces the overwintering population. The survivors produce their first young, which look and act like miniature adults, in late May. The progeny grow and mature quickly, releasing more young in mid-July—just when the shorebirds are beginning to arrive.

The females are the first arrivals. During the short Arctic breeding season, they produce a clutch of three or four eggs. The precocial chicks are capable of taking care of themselves at birth, and the females leave shortly thereafter, first flying from their central and eastern Arctic breeding grounds to the west coast of James Bay, where extensive marshes and mudflats provide an ideal, food-rich staging area for their 1,500-kilometer (932-mile) flight to the Maritimes. The females arrive in Fundy with a fat reserve and, unhampered by the more aggressive males, immediately begin topping up for the last leg of their epic journey by selectively removing the larger, overwintering mud shrimp.

*following spread:* A blizzard of migratory shorebirds converges on the Bay of Fundy's inner basins each summer.

By late July the males and juveniles, who appear to travel together, begin arriving, in time to take advantage of the second generation of mud shrimp production. The birds consume between 9,600 and 23,000 mud shrimp during a single 12.5-hour tidal cycle. To maximize their food intake, the birds must follow the retreating tide very closely. As the tide falls the mud shrimp emerge from their burrows to feed on algae or to seek a prospective mate. They do so for only about twenty minutes, when the mudflats begin to dry out, forcing the mud shrimp to retreat to their burrows. The ever-vigilant peeps, feeding with an almost maniacal diligence, are able to double their weight, from 20 to 40 grams, in ten to fourteen days of feeding. These are the highest weight gains for shorebirds recorded on the Atlantic coast, suggesting that there is a selective advantage for birds using Fundy, which shorebird biologist Peter Hicklin has cogently described as a "fat station."

The shorebirds need every bit of this caloric capital for their 4,000-kilometer (2,500-mile) journey over open water. They time their departure in advance of a cold front, signaled by a northwest wind, which carries them

in the direction of Africa, seemingly off course. But over the Caribbean, they encounter the northeasterly trade winds, which push them back toward their desired destination of the north coast of South America.

Flying at an average airspeed of 60 kilometers (37 miles) per hour, they can complete their epic journey in forty to sixty hours. Once they commit themselves by taking a ride on the northwesterly offshore wind, there is no turning back. Neither can they rest on water, since they do not have the salt-water-clearing mechanisms that seabirds do. The epic journey is a nonstop ordeal, and it exacts a tremendous toll. Birds arrive in South America with their fuel reserve of fat nearly gone, and some birds actually deplete muscle tissue to complete their journey.

Shorebirds are generally capable of storing enough fat to make flights of up to 4,600 kilometers (2,900) miles, and it is approximately 4,000 kilometers (2,500 miles) from the Bay of Fundy to Suriname, where two-thirds of the semipalmated sandpipers winter. Clearly, the Bay of Fundy plays a critical role in the survival of this species. For this reason, the shrimp-rich mudflats of Fundy in Shepody Bay and the Bight area of Minas Basin have been given protection as Western Hemisphere Shorebird Reserves. This is a network of sites in North and South America, including Delaware Bay, that are deemed vital to the successful completion of the shorebirds' life cycles. They are links in a chain, with the loss of any one posing a serious threat to the survival of a species. The inner Bay of Fundy is such a link for semipalmated sandpipers.

Peregrine falcons are now frequently seen diving through Fundy's massive shorebird flocks, taking their prey on the wing. They are the fastest bird of prey, attaining speeds of 350 kilometers (217 miles) per hour as they make their stooping dives. Peregrine falcons are endangered throughout their range in eastern North America, the population having suffered a precipitous crash after 1950, when widespread use of the pesticide DDT caused eggshell thinning and thus breeding failure. Peregrines disappeared from the Bay of Fundy in the late 1940s after DDT was sprayed heavily on budworm-infested forests.

Peregrines have been successfully reintroduced to the inner Bay of Fundy, however, through a ten-year captive-release program begun in 1982 at Fundy National Park. Just as the arrival of shorebirds is timed to the explosion of mud shrimp in the Fundy mudflats, the young peregrines' fledging coincides with the shorebird migration into the region. The peeps serve as a ready source of prey for the young birds as they learn to hunt for themselves.

Peregrine falcons have made a remarkable recovery from DDT poisoning and now occupy most of the traditional aeries in the Bay of Fundy.

THE CANADIAN PIONEER marine biologist A.G. Huntsman listed the elemental factors that were important to the great productivity observed in the Bay of Fundy: ". . . a thin layer of water, containing a variety of substances in solution, lying between the air above with its gaseous substances and the earth below with its various solid substances, and exposed to the sunlight coming through the transparent air, seems an ideal location for the production of life. Shallow inlets from the ocean have such characters, and one of these is the Bay of Fundy." These critical conditions are met not only at the mouth of the bay, where Huntsman concentrated his research, but elsewhere along the seaweed-draped shoreline and islands of the Gulf of Maine, as well as on Georges Bank, the shallow "oceanic miracle" at the center of gulf. Together, these ecosystems account for the great diversity of life encountered in the tidally driven Bay of Fundy–Gulf of Maine ecoregion.

# 6

# RIVER INTO THE SEA

## *The Gulf of St. Lawrence*

THE FIRST DESCRIPTION of the Gulf of St. Lawrence was Jacques Cartier's chronicle of his 1534 and 1535 voyages into the gulf and river, which bear the same name. He writes in wonder of the Bird Rocks, barren, high-sided islets at the center of the Gulf, "completely covered with birds, which nest there, as a field is covered with grass." The birds were gannets, murres, and the now extinct great auks. On nearby Brion Island, he marvels at his first encounter with walruses, which he colorfully describes as "great beasts, like large oxen, which have two tusks in their jaw like elephant's tusks and swim about in the water."

During these two voyages, Cartier explored the many arms of the Gulf, which the 20th-century natural historian Wayland Drew has aptly described as being roughly starfish-shaped. The longest arm reaches into the interior of the continent through the St. Lawrence River. A second arm points north toward Labrador through the Strait of Belle Isle. A shorter arm curls under the Gaspé Peninsula to create the Baie des Chaleurs, and another long arm cradles the crescent of Prince Edward Island to become the Northumberland Strait. Finally, a fifth arm reaches toward the open Atlantic through the Cabot Strait, between Newfoundland and Cape Breton Island.

The Gulf of St. Lawrence is a place of contradictions: in summer it boasts the warmest waters north of the Carolinas, but in winter it is covered with ice for three months. The gulf

‹ The Gulf of St. Lawrence has been called a giant estuary, a place where freshwater meets salt.

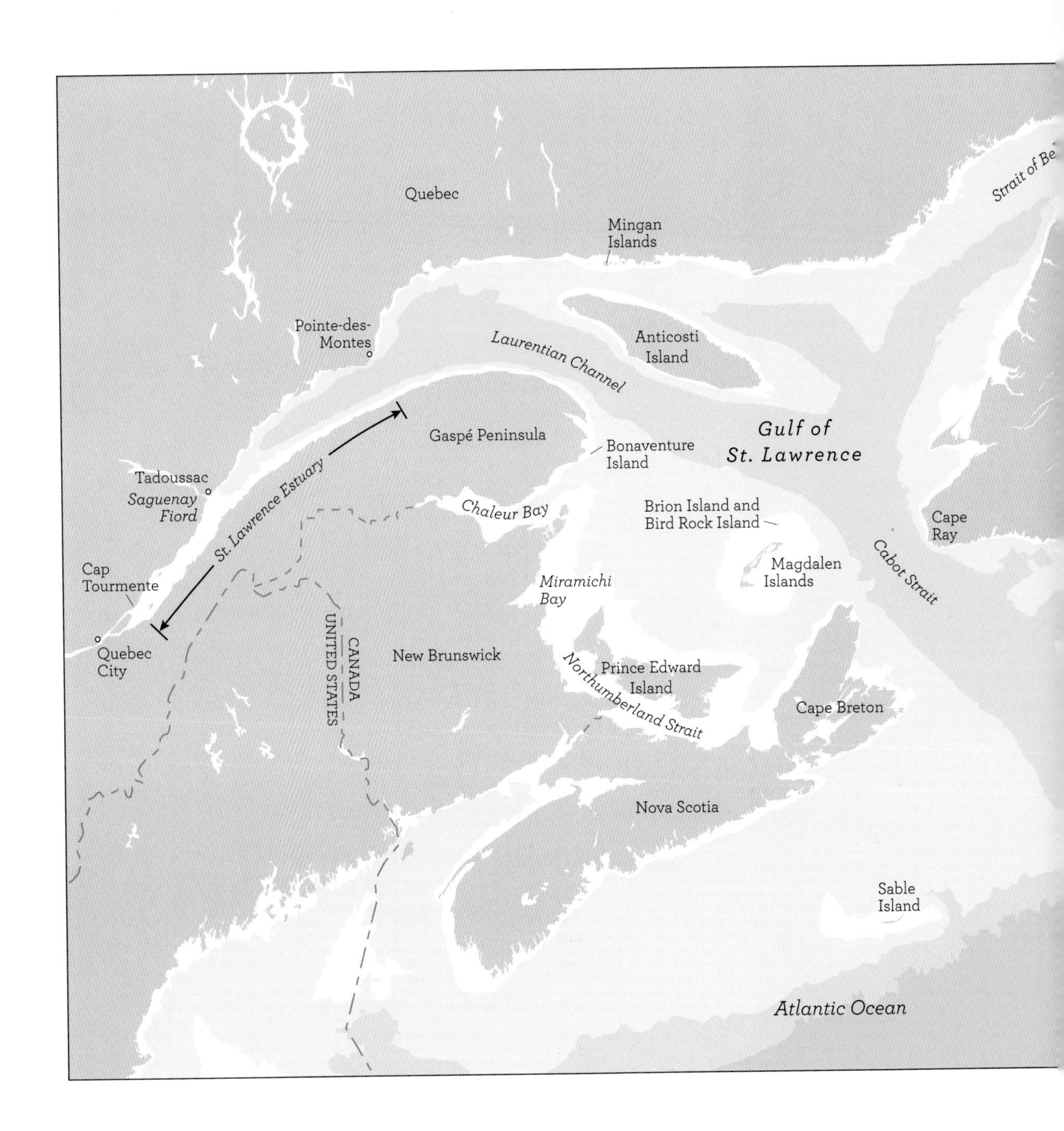


MAP: The St. Lawrence Estuary and Gulf of St. Lawrence

is also difficult to define as an oceanographic entity. Oceanographers have pondered whether it acts more like an inland sea or like a giant estuary; in fact, it appears to be a bit of both. On the one hand, it is a huge estuary—a transition zone where freshwater and tidewater meet and mix and which exhibits some characteristics of the land-based world of freshwater and some of the salt waters of the marine zone. On the other hand, the gulf is a semi-enclosed inland sea, which opens to the North Atlantic through the Cabot Strait and Strait of Belle Isle, making it subject to tidal, oceanic influence.

Two features set the Gulf of St. Lawrence apart from other semi-enclosed seas, such as the Baltic Sea and the Black Sea. Normally an estuary discharges its freshwater flow directly into the deep waters of the continental shelf, but in the Gulf of St. Lawrence the freshwater is discharged into a shallower basin, a drowned lowland formed during the Carboniferous period, when the major channels that crisscross it were first eroded. The second distinguishing feature is the enormous size of the drainage basin that feeds it, primarily through the St. Lawrence River. This source of freshwater includes the Great Lakes and, at 226,000 square kilometers (87,000 square miles), is nine times as large as the surface area of the gulf itself. Its massive freshwater flow is greater than the runoff from the entire east coast of the United States, from Canada to southern Florida, and its influence is felt as far away as the Scotian Shelf and the Gulf of Maine, a distance of 1,300 kilometers (800 miles), where catches of some twelve species of fish landed from the Gulf of Maine and Georges Bank can be correlated with river outflows.

But salt water—shelf water—enters the mouth of the great river through the Laurentian Channel, which cuts diagonally across the gulf. This channel exceeds depths of 350 meters (1,150 feet) along most of its length, dipping to more than 500 meters (1,640 feet) near the Cabot Strait and extending 250 kilometers (155 miles) across the continental shelf. Despite the enormous flush of freshwater from the St. Lawrence River, it is swallowed up and overwhelmed by the tidally driven Gaspé Current, which entrains more than a hundred times the volume of the river runoff in its counterclockwise circulation of the gulf.

The salt and fresh waters of the gulf perform a kind of oceanic pas de deux. The fresher, lighter water from the river moves on the surface toward the sea, while the heavier, saltier ocean water moves, at depth, toward the river. These two masses are on a collision course and begin to mix just downriver from Quebec City, but the major mixing occurs where the Saguenay River empties into the St. Lawrence. Between Quebec City and Tadoussac, the water is brackish—a mixture of fresh and salt—and also turbid, since intense mixing brings the sediments into the water column. This area is defined as the Upper Estuary, and the Lower Estuary extends seaward from Tadoussac to Pointe-des-Montes on Quebec's North Shore.

*Waterfowl and White Whales*

Each spring and fall, cacophonous flocks of greater snow geese descend on the Upper Estuary. In the shadow of the glacial-scoured granite mass of Cap Tourmente, the marshes along the North Shore of the St. Lawrence are turned white as if by a premature snowfall by the massing of greater snow geese during their migration from their breeding grounds in the Canadian Arctic to their wintering areas along the Mid-Atlantic coast. During the fall, the entire world population of the birds fatten on the rhizomes of American bulrush, which flourishes in the vast expanses of tidal marshes between Quebec City and Montmagny. As many as 800,000 geese gather at the Cap Tourmente National Wildlife Area alone. (The protected areas along the river have been critical to the revival of this species, whose numbers were reduced to three thousand at the turn of the century).

The birds begin arriving when the first frosts in the Arctic force them south. While foraging, the geese follow the tide to take advantage of the softer sediment, using their strong bills to dig into the marsh muds. They strip the tidal marshes bare of the bulrush before their departure in October. This feeding activity may help the growth of wild rice, which in turn enhances the attractiveness of the marshes for other waterfowl, such as dabbling ducks. The Montmagny area is a traditional fall staging area for northern pintails, and Cap Tourmente attracts pintails, green-winged teal, and American black ducks.

On the fat reserves acquired here, some snow geese fly the 900 kilometers (560 miles) nonstop to marshlands along the eastern seaboard from New Jersey to South Carolina. They concentrate on coastal salt marshes, where they feed on smooth cordgrass and salt marsh hay. The rapidly expanding population of snow geese has caused "eat-outs" of the southern marshes, which have

A field of greater snow geese lifts off near Cap Tourmente, Quebec.

had a number of detrimental effects on coastal salt marsh ecosystems, including increased erosion and decreased primary productivity. Grazed marshes in New Jersey have also experienced a decline in the number of clapper rails as a result of damage to their nesting sites. Other typical salt marsh invertebrates, such as salt marsh snail, ribbed mussel, and fiddler crab have also declined, and black duck populations may have been affected as a result.

The over-grazing opens up the monotypic *Spartina* stands, however, leading to a greater diversity of plant species and providing habitat for shorebirds and wading birds. The feeding activity also helps to recycle nutrients in the system for other salt marsh consumers. The geese return to the St. Lawrence marshes in the spring, though they tend to spread out more extensively along the river and estuary at this time.

SNOW GEESE ARE not the only waterfowl to exploit the river's estuarine environment. Between Rivière-du-Loup and Kamouraska, in the Lower Estuary, a series of dramatic bald islands stand in stark profile against the river's steely face and, beyond that, the ramparts of the Canadian Shield on its North Shore.

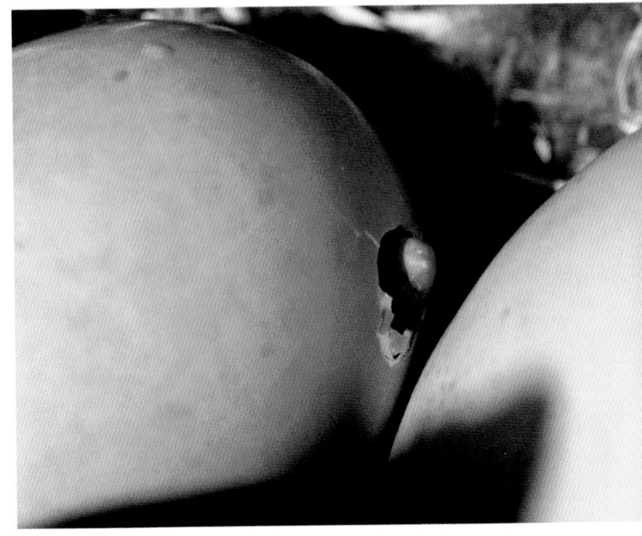

An eider duckling (left) rests warmly among the down plucked from its mother's breast, and a newborn pips its egg (right).

I once had the privilege of visiting these islands to witness the annual gathering of eiderdown, the profits from which are used to support conservation of this island ecosystem.

These islands are the stronghold of the eider duck in North America. Some 25,000 pairs breed on the more than twenty islands that stem the tide in the Lower Estuary, and a further 5,500 breed on the shores and islands of the gulf itself. Although these numbers are impressive, they were once far greater if the historical reports are to be believed. When the pioneering bird artist John James Audubon visited Île Biquette in 1833, he wrote in his journal: "So abundant were the nests of these birds on the islands that a boat load of their eggs might be collected if they had been fresh; they are then excellent for eating."

In fact, the excellent culinary qualities of eider eggs almost led to the species' demise. Egging of these islands seriously depleted the eider population, and hunting also had an impact, since eiders are the largest of North America's sea ducks and are therefore also highly prized for the table. It took the passage of the Migratory Birds Convention Act of 1917 to slow the depredation and carnage. The breeding numbers have since rebounded in the estuary, where nearly 90 percent of the breeding birds are found on five islands: Île Biquette, Île aux Fraises, Île Blanche, Île aux Pommes, and Îles du Pot-à-l'Eau-de-Vie. Of these, Biquette harbors the greatest number of nests—some seven thousand in total.

The birds form dense breeding colonies, with more than four hundred nests per hectare. Eiders begin laying in late April, and nearly all nests have eggs by June. The female constructs her nest on the ground from grass and seaweed and lines it with large amounts of down feathers, which she plucks from her breast. (The bird's scientific name reflects the qualities of its down: *Somateria* meaning "body wool," and *mollissima*, "very soft".) She lays four to six large, olive-colored eggs, which are kept warm by the super-insulating properties of the down and her own body. The brooding female leaves the nest only for a few minutes at a time while she incubates the eggs, and during these four weeks she does not eat. After breeding season, the promiscuous males do not take part in the incubating or rearing of the young, but instead return to the water, their handsome black and white markings creating a dramatic sight wherever the birds gather off the coast in large "rafts."

The precocial ducklings leave their nests within hours of hatching and with their mothers make immediately for water. It is a perilous first journey, for they must run a gauntlet of predatory herring and great black-backed gulls, which often nest on the same islands. It is, as nature writer Franklin Russell once observed, "a mathematical experiment in survival"—an experiment in which the odds are stacked heavily against the duckling. A study on Île aux Pommes revealed that 63.3 percent of nests were destroyed by predation, 14.1 percent were abandoned, and only 22.3 percent of nests produced one or more young. On Île aux Pommes I had watched as a newly hatched chick fell behind its mother when it was briefly rebuffed by waves lapping the shore. With great exertion it struggled to the top of a beach stone; then, as it paused momentarily, a herring gull swooped in, plucked the duckling cleanly from the rock, and with one devastatingly efficient toss of the head swallowed it whole.

Such a scene calls into question why eiders choose to nest among predatory gulls. Female eiders are large enough to defend themselves against gulls and rarely leave their nests; therefore predation on eggs is low. In some instances, the wary gulls might perform a positive function by warning eiders of a potential threat, such as a fox or a human. The study on Île aux Pommes showed that eiders nesting within 5 meters (16 feet) of herring gull nests actually had higher success, indicating that territorial defense by herring gulls does confer an advantage on the eiders.

Once in the water, the adults and ducklings form extended families called crèches—literally nurseries. Chick-crèches are typical of colonial nesters like eiders, whose eggs hatch roughly at the same time. Such groupings range in size, from a single duckling with an adult female other than its own mother, to huge creches of fifty or more ducklings and several accompanying females, often nonbreeders whimsically referred to as "aunts."

In the past, it was thought that crèches constituted a deliberate cooperative system to care for the young or were an evolved behavioral trait. But recent research indicates that crèches form accidentally, when broods run into each other on the crowded islands, or spontaneously, in the face of gull attacks—the duck equivalent of circling the wagons. Regardless, the larger the crèche is, the lower the duckling mortality, a reflection of the dilution principle whereby the fledgling loses itself in the crowd and lowers its individual risk of becoming a victim of predation.

These protective extended family units stay together for the entire rearing period of ten weeks. Once the females guide the ducklings to shallower, more protected nearshore waters, the young are able to feed themselves by diving for molluscs and crustaceans. Following the nesting season, the eiders move downstream, males and nonbreeding females converging at sites in the Lower Estuary where their chief prey item, the blue mussel, is most abundant. Ducklings and the adult females accompanying them gather along the south shore of the St. Lawrence, during July and August, in areas where the brood has access to its principal prey, the periwinkle. Eiders swallow their prey whole, shell and all, to be ground up in the bird's muscular gizzard.

The St. Lawrence eiders begin to move south in mid-September, flying low over the water in a long-line formation, stringing out impressively along the horizon and appearing to skim and dip between the waves. They show up along the northeast coast of New Brunswick from October through November, when they are forced out of the gulf by ice formation. Some eiders, however, remain in the gulf, where ice conditions allow access to food. Aerial surveys have shown that roughly 30 percent of the eider population overwintering along the coasts of eastern North America is found in the Gulf of St. Lawrence, with 95 percent of that total concentrated along Quebec's North Shore, between Anticosti and the Mingan Islands.

Those that do migrate take two routes: many pass overland across the Chignecto Isthmus from the Northumberland Strait to the Bay of Fundy, and the balance migrate through the Strait of Canso. While migrating over land, they assume a higher flying pattern—perhaps to allow a view of water on the other side—with the front ranks massed in a semi-circle and followers in a long wavy line behind. Many are destined to spend the winter in Massachusetts Bay, though smaller numbers overwinter in ice-free areas along Nova Scotia's south shore, forming large rafts in winter, with as many as a thousand individuals.

BESIDES ITS BREEDING islands and food-rich fringing marshes, the St. Lawrence River itself is a wildlife thoroughfare. While on the eider islands, I witnessed a remarkable parade—a large pod of beluga whales, perhaps as many as a hundred animals, passing by on their way upstream. Like synchronized swimmers, their startlingly white bodies "porpoised" out of the water in unison, and their bulbous heads spy-hopped above the river's blue waters. This pod of adults and accompanying young constituted a significant portion of the world's most southerly population of belugas, which are primarily an Arctic species.

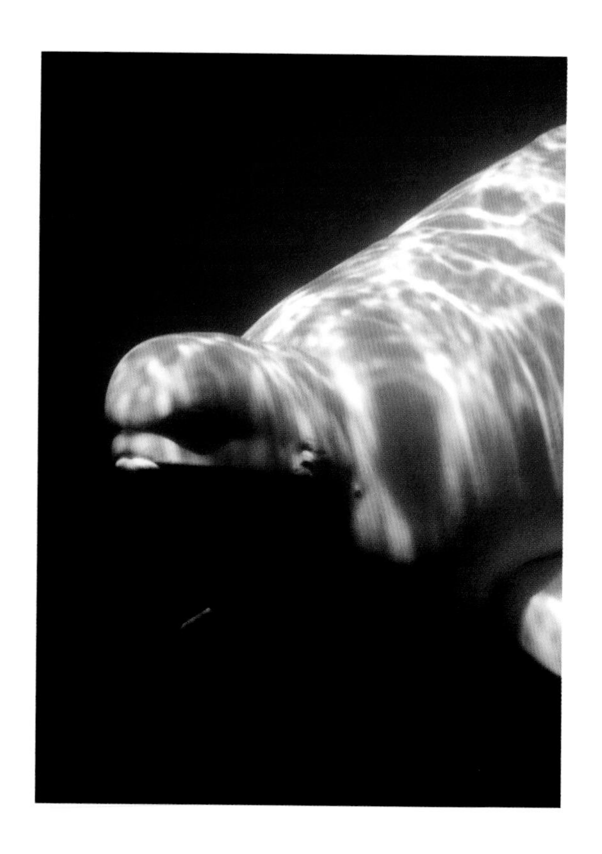

Alone among whales, belugas have what appears to be a "neck" and a high degree of lateral head movement.

Belugas belong to the suborder of toothed whales, the Odontoceti, and their closest relative is the narwhal. Beluga means "the white one" in Russian, and their whiteness marks them as an Arctic species, like many other northern mammals and birds, including greater snow geese. They are well suited to Arctic conditions, having a thick skin and blubber layer to insulate them from the cold, and lacking a dorsal fin, which may be an adaptation to swimming around and under ice.

Besides their creamy white coloring, belugas have other distinctive anatomical and behavioral characteristics. They have a noticeable constriction behind the head, which can best be described as a "neck." Because the neck vertebrae are not fused together, belugas, alone among whales, have an unusual degree of lateral head movement. They have a bulbous forehead—the "melon"—whose shape can be changed by muscular contractions and is thought to act as an acoustic lens, enabling belugas to focus their echolocation clicks directionally. This sonar capability helps them navigate in shallow waters and find prey, especially in dark and turbid environments such as the St. Lawrence estuary. Belugas have a narrow transmitted-beam pattern that is useful for navigating under ice and finding air pockets under an ice pack, which seem to be adaptations to an Arctic acoustic environment.

Belugas have been described as the noisiest of whales, earning them the sobriquet "sea canaries." Recordings of their vocalizations made in the Saguenay Fiord were described as "high pitched resonant whistles and squeals, varied with ticking and clucking sounds, slightly reminiscent of a

The limestone base of Anticosti Island preserves a wealth of fossil organisms that inhabited a warm sea 60 million years ago.

string orchestra tuning up, as well as mewing and occasional chirps." Finally, belugas have a pair of flexible lips that aid in the consumption of bottom-dwelling organisms such as flounder, eels, and crustaceans, but, to the anthropocentric-minded, lend a smiling expression to their faces. Mostly, however, they feed on small schooling fishes, such as sand lance, capelin, herring, and smelt, that frequent the estuary.

Belugas are unique among whales in that they are tied to estuarine environments. In Arctic waters they return to the same estuary year after year, even in the face of hunting pressure. They often give birth in the fresher waters; in the case of the St. Lawrence population, the birthing territory is the Saguenay estuary. The freshwater environment also stimulates accelerated growth of the epidermis, or molting.

Although belugas are a circumpolar species, found in all the northern seas, they have been in the St. Lawrence environment for a very long time, as demonstrated by the discovery of bones in marine deposits of the ancient Champlain Sea dated to 10,500 years ago. The whales were hunted by the indigenous Montagnais Indians, but large-scale exploitation began with European settlement. Brush weirs, first erected at Kamouraska at the beginning of the 18th century, proved deadly. Five hundred belugas were captured on a single tide at Rivière-Ouelle in 1880, and eighteen hundred in one season at this same site. At the time, the population was estimated at five thousand animals. As late as the 1930s, a bounty was placed on belugas (which were erroneously thought to be depleting commercial fish stocks such as cod and salmon), leading to the death of nearly two thousand whales. Although intensive hunting was banned in the 1950s, the population was officially protected only in 1979; by then, however, it had been reduced to five hundred animals, probably one-tenth of historical levels. Current estimates place the population at a thousand animals.

The slow recovery of the species has been attributed to several factors. Their range has contracted in recent decades because of the loss of critical habitat. Hydroelectric development on the Bersimis, Manicouagan, and

## THE SOUTHERNMOST FIORD IS AN ARCTIC ENCLAVE

THE 170-KILOMETER-LONG (105-mile) Saguenay Fiord was carved into the Canadian Shield by glaciers making their way to the sea. In the process, it gouged out a classic steep-sided, U-shaped valley. Like most fiords created in this way, the Saguenay has a deep inner basin and a shallow sill at its mouth. But it is also one of the southernmost fiords in the world with a unique fauna. Many of the benthic invertebrates found in the colder deep-water basin within the fiord are typically Arctic species and are not found in either the St. Lawrence River or the gulf. Likewise, of the fifty fish species, more than 80 percent have an Arctic affinity, and many are also endemic to the fiord. For these reasons, the fiord is considered by many biologists to be a biogeographical Arctic enclave—an isolated Arctic world unto itself.

The Saguenay is also unusual in that it empties into an estuary rather than the sea. As a result, three water masses collide and tides mix the cold, salty waters of the gulf, at the head of the Laurentian Channel, with the freshwaters of the two river systems that meet here, bringing nutrients and oxygen from the bottom into the light zone, boosting productivity, which is higher than that observed in the St. Lawrence estuary. This upwelling of cold waters creates Arctic-like conditions, which accounts for the mouth of the Saguenay being the major habitat of the resident beluga population.

The steep-sided Saguenay Fiord shelters many Arctic species, including a resident beluga population.

Outardes Rivers has altered the temperatures and biological productivity in their estuaries, which once furnished one-third to one-half of the species' summer habitat.

Today, belugas are largely confined to the confluence of the Saguenay and St. Lawrence, where local upwelling creates ice-free waters in winter and near Arctic conditions in summer conducive to belugas. Some belugas disperse as far as the limits of the estuary environment at the mouth of the St. Lawrence. Formerly, they also congregated along the coast of the Gaspé in spring and the lower North Shore of the Gulf in summer, but they now appear at these sites only sporadically.

Chemical contamination may be another factor limiting population growth. The St. Lawrence beluga habitat is downstream of the highly populated and industrialized Great Lakes watershed. Autopsies of stranded individuals have detected a veritable cocktail of contaminants, including high levels of fifteen toxic chemicals and nine heavy metals—in some cases, twenty-five to one hundred times the levels found in Arctic populations. Among the pollutants that have been found are PCBs and PAHs, DDT and other pesticides, and the heavy metals cadmium, mercury, cobalt, and lead. The source of these contaminants seems to be the whales' prey, including the American eel, which migrates downstream from the industrialized regions of the Great Lakes and the upper St. Lawrence on its way to breeding grounds in the Sargasso Sea. The stranded whales also had unusually large numbers and types of lesions, ulcers, and tumours, which, if they did not cause death, might have affected the belugas' overall health and reproductive capacity.

The natural rate of recruitment of St. Lawrence belugas is only half that of the other populations in Arctic waters, and it could be that the population of St. Lawrence belugas had already fallen below a critical level to allow it to compensate for the natural rate of mortality. Now chemical pollution, loss of habitat, and accidental deaths due to high traffic levels on the river may well be exacerbating an already perilous condition.

*North Shore Blues and Daughters of the Sea*

Jacques Cartier's description of the North Shore of the Gulf of St. Lawrence, which he reconnoitered on his first voyage to the New World, is anything but charitable. "If the soil were as good as the harbours, it would be a blessing," he writes, "but the land should not be called the New Land, being composed of stones and horrible rugged rocks; for along the whole of the north shore [of the gulf] I did not see one cart-load of earth and yet I landed in many places . . . I am rather inclined to believe that this is the land God gave to Cain."

Waves roll in along the Gulf of St. Lawrence's rugged North Shore.

Quebec's rugged North Shore is deeply indented with many bays and inlets and the estuaries of brash rivers such as the Moise that spill from the Precambrian Shield and fan out to form large deltas. The cold Labrador Current enters through the Strait of Belle Isle and imparts an Arctic character to the waters, which is also reflected in the flora and fauna on land. Cold water fishes, such as Atlantic herring, capelin, Atlantic cod, Arctic cod, Greenland halibut, and redfish, are common, and Atlantic salmon, brook trout, and rainbow smelt run up the many rivers. Juvenile fishes are dominated by capelin, an important food item for the 200,000 seabirds that nest in small colonies on the North Shore cliffs, including a disjunct colony of Caspian terns at Île à la Brume on the lower North Shore.

In recent decades, researchers have discovered that not only beluga whales frequent this coast. At least twenty species of cetaceans have been observed in the region during the ice-free months, including baleen species—fin, minke, humpback, and blue whales—as well as toothed whales such as harbor porpoises, Atlantic white-sided dolphins, and white-beaked dolphins. The area around the Mingan Archipelago seems to be particularly important to blue whales. Studies carried out by the Mingan Island Cetacean Study group have demonstrated that individual whales regularly return to the area between the Mingans and the Saguenay River.

NORTHERN KRILL

Blue whales are the largest animals ever known to have lived on Earth, reaching lengths of 30 meters (100 feet) and weighing as much as 75 tonnes. They are found in all of the world's oceans and migrate seasonally, usually moving toward the poles in the spring to take advantage of the high zooplankton production there in summer. In some cases they may form relatively discrete feeding populations, migrating together from one area of high productivity to another. Although there is some evidence suggesting that the Davis Strait and Gulf of St. Lawrence whales belong to the same population, acoustic tracking of blue whales indicates that they may range over the entire North Atlantic ocean basin and therefore may comprise a single stock, a question only genetic analysis may be able to answer.

What we do know is that blue whales are present in the Gulf of St. Lawrence for most of the year. They are forced to leave in early winter, however, to avoid entrapment in ice and do not return until ice breakup in the spring. There are two peak periods for their presence along the North Shore: one in early April to early June, and another from August until late October. They are very nomadic and rarely spend more than ten days feeding in a particular area. In addition to Quebec's North Shore, they have been sighted in areas off the Gaspé Peninsula, around Anticosti Island, and in the St. Lawrence River estuary as far upriver as Tadoussac.

Stomach content analysis has shown that the food of blue whales in the North Atlantic consists entirely of krill, relatively large euphausiid crustaceans. Two species, *Thysanoessa raschii* and *Meganyctiphanes norvegica*, are particularly important food sources in the gulf. During the daylight hours, blue whales are frequently observed feeding along the North Shore's 100-meter (300-foot) contour, where krill are concentrated.

Fortunately for blue whales, humans do not compete with them for their prime food source, which has no commercial value. The great whale's numbers were severely depleted by whaling, however, especially in the second half of the 19th century, when steam-powered vessels carrying deck-mounted harpoon canons came into wide use. Blue whales in the North Atlantic were finally given full protection in 1955, but by then the population had been decimated. It has been estimated that perhaps as many as fifteen hundred blue whales were found in eastern Canadian waters when modern whaling began. (Some studies have suggested that as many as fifteen thousand may have ranged the entire North Atlantic in the pre-whaling period.) By the early 1960s, their numbers were in the very low hundreds at most. By the mid-1990s, some 350 individuals had been photo-identified in Canadian and New England waters, of which 320 were identified in the Gulf of St. Lawrence. But in nineteen seasons of observation, only nine blue whale calves were observed along the North Shore. A number of explanations might account for their low presence: it could be that lactating females prefer other areas, or that weaning occurs before the whales arrive along the North Shore, or—the worst-case scenario—that few calves are being produced in the population, for reasons that are not well understood.

THE MINGAN ARCHIPELAGO consists of fifteen islands and forty or more islets that stretch for 80 kilometers (50 miles) between the North Shore of Quebec and Anticosti Island, in the Jacques Cartier Channel. Many have been sculpted by water and ice into limestone monoliths, or flower pots, which the great Quebec botanist Brother Marie-Victorin colorfully described as "daughters of the sea." The tides, currents, and ice continue to sculpt them into ever-changing shapes.

The islands' steep cliffs and their isolation make them attractive to breeding seabirds, including gulls, terns, eiders, and 35,000 pairs of Atlantic puffins. However, it is the plants that are the most remarkable biological feature of the islands.

Because of its rich limestone soils, the archipelago boasts much greater plant diversity than the much larger area of the adjacent North Shore—450 species of vascular plants as compared with 380 species. Over 100 of these are classed as species of interest because of their relative rarity. One group of plants that have adapted to the highly alkaline soils is the saxifrages. The name means "rock breakers," and the plants anchor themselves in the smallest cracks in the cliff faces, where in June they put forth a profusion of yellow, white, and fuchsia blossoms.

Perhaps the hardiest island habitat is the wind-swept barrens, which hosts more than thirty varieties of vascular plants native to arctic and alpine regions, including alpine chickweed and entire-leaved mountain avens. Farther offshore lies the great limestone island of Anticosti, a juggernaut sitting astride the mouth of the St. Lawrence. Its Dover-white cliffs tower above the gulf, and among the natural wonders of its interior is Vaureal Falls, which is higher than Niagara and has carved a spectacular canyon through the limestone karst. It

A pair of gannets engage in an elaborate mating ritual, which includes sky pointing and bill fencing.

also boasts twenty-four rivers, which are host to Atlantic salmon. The most famous is the Jupiter, the color of jade as it courses over its limestone bed. The island is a geological relic of a sea that covered the area 60 million years ago. Today, it is littered with marine fossils and some six hundred types, including trilobites, have been documented. Its limestone soils now support an unusual, acid-rain-resistant flora, including nineteen species of orchids, the rarest of which is the bog plant *Orchis rotundifolia*.

*Gannets and the Gaspé Current*

The Gaspé Current sweeps around the peninsula of the same name as part of the overall counterclockwise, estuarine circulation within the gulf. In doing so, it picks up the downstream flow of plankton from the Lower Estuary. High levels of nutrients from the tidal forces and from the estuary support a community of large diatoms during the summer, which in turn are fed upon by large species of zooplankton, such as *Calanus* species and euphausiids. This biological productivity moves up the food chain through fish and seabirds, in particular to mackerel and gannets.

A quarter of the 400,000 seabirds breeding in the gulf do so on the cliffs and islands off the eastern Gaspé Peninsula. The most important site is Bonaventure Island, adjacent to the iconic Percé Rock, the 100-meter-high (300-foot) limestone island whose soft sediments have been dramatically pierced by tidal action. Bonaventure is the largest colony of northern gannets in the North Atlantic, now boasting 122,000 pairs. It is also the most accessible, as visitors can approach within feet of the seabirds as they court and nest on the 80-meter-high (260-foot) cliffs. There are few spectacles in nature as impressive as the massing of these magnificent white birds on their nesting grounds. As biologist Bryan Nelson observed in *The Gannet*: "A seabird colony is, in a sense, a supra-organism . . . not just a homogenous mass of birds."

Gannets are large seabirds with wingspans of almost 2 meters (6.5 feet). Their rudimentary nests are made from a random collection of seaweed, sticks, and moss and are occupied year after year, with accumulated droppings, fish skeletons, and feathers adding to their bulk. Gannets always nest in very dense colonies, even, it seems, where there is a surplus of nesting space. Nests are packed as closely together as possible, perhaps as an anti-predator device, but more likely to provide social stimulation, which leads to reproductive success.

The timing of breeding appears to be especially important to gannets breeding in the northwest Atlantic. Gannets arrive en masse at Bonaventure, and there is fierce competition—fighting—for a site among, or near, established birds. The courtship antics of pairs are elaborate, even comical, involving much head bobbing, sky-pointing, and fencing with their bills, but the fighting and wooing are thought to stimulate neighboring birds to breed. The earlier the better seems to be the rule, as young grow more rapidly early in the season and early fledgers are more likely to survive. In northern climes, such as on the Gaspé, it is also important that the breeding cycle be completed before late September, when the weather usually deteriorates rapidly. Gannets lay a single egg beginning in late April or early May, and by the first week of July the young gannets begin to hatch. Parents will feed the young with regurgitated fish until they fledge in September.

Land, bird, and sea come together dramatically on Bonaventure Island, the largest gannetry in the North Atlantic.

Gannets are ungainly on land because of their short legs and large webbed feet. In the air, however, they are supremely graceful, using their magnificent wingspan to glide over the wave tops, sometimes for hours, with only the occasional beating of their wings. This tactic, called "wavehopping," employs the natural updraft off each wave crest. It is in their diving and fishing behavior, though, that gannets truly excel. Folding its wings and diving almost vertically, a gannet seems to drop from the sky like a lead plumb bob. Upon impact, spray sometimes spouts 3 meters (9 feet) into the air. The gannet is particularly well adapted for this daredevil behavior. Unlike most birds, it has binocular vision (the eyes are positioned so that both see forward simultaneously), allowing it to estimate how far below the surface the prey lies. It has no nostrils, and its upper and lower bill fit tightly together to prevent ingestion of water on impact. Upon diving, the gannet inflates air cells located between the skin and shoulders and the muscles beneath. These cells act as shock absorbers to cushion its body from the tremendous force of impact. The dive drives the bird below its prey; it then swims up with its wings and webbed feet to capture the fish. The great gannet colony on Bonaventure Island is sustained by mackerel, which migrate into the gulf to spawn at the same time that the young are hatching and the colony's nutritional needs are greatest.

### *Sea Meadows, Lagoons, and Dunes*

Once around the Gaspé, you enter the Magdalen Shallows, an underwater plateau some 150 kilometers (93 miles) wide, cut here and there by several deeper troughs. Generally, the waters are, as the name would suggest, shallow—some 80 meters (260 feet) deep, though, of course, much shallower near the shore. The Gaspé Current conveys freshwater from the St. Lawrence estuary, which, combined with the shallow depth, results in the warmest waters in Eastern Canada. These shallow waters are also the most productive in the gulf region.

Near my home on the Northumberland Strait is Baie Verte, a bay named for the luxuriant salt marshes that line its shores. Eelgrass also grows in abundance in its shallow waters, further boosting the primary productivity, which in turn nourishes rich beds of invertebrates. Legions of clam-diggers spread out on the mudflats to search for quahogs and razor clams, and wildlife also congregate to reap the marine harvest. From my vantage point at the Tidnish Dock, on any given night in late summer I can watch as two dozen great blue herons string out along the water line to pluck fish from the shallows. Near

The shallow southern gulf supports large numbers of fish-eating birds like these cormorants.

the shore, greater yellowlegs and willets probe the mudflats for invertebrates as a small flock of sandpipers wheels by. Common terns from a nearby colony enliven the air with their constant chatter as they dive for silvery prizes. Gray and harbor seals loaf on the rocky reefs, while a dark flock of double-crested cormorants spread their wings to dry them, creating eerie silhouettes.

Scenes such as this confirm the observation made by Nicolas Denys in 1671, when he called the southern gulf "a land of greatest abundance" in his *Description and Natural History of the Coasts of North America*. The coast throughout the region is distinguished by barrier islands, large beaches, dunes, estuaries, salt marshes, and lagoons—a mosaic of diverse and productive habitats that support large nesting colonies of waterbirds, which feed on the small fishes found in the nearshore areas. The shallow waters that surround Prince Edward Island, for example, support two major populations of fish-eating birds: a continentally significant population of breeding great blue herons as well as the largest colony of double-crested cormorants in eastern North America, numbering twelve thousand pairs. The beaches also support the largest population of piping plover in Eastern Canada.

Great blue herons stalk through the shallows, evoking Dylan Thomas's phrase, "the heron-priested shore."

The southern gulf, seen from a bird's-eye view, consists of a system of barrier islands, beaches, and dunes that stretch along New Brunswick's low-lying shoreline, as well as the north shore of Prince Edward Island and the northern shore of the Nova Scotia mainland and western Cape Breton Island.

The sand for these islands, beaches, and spits is furnished from the soft sandstone underlying the region and redistributed by the forces of tides, winds, waves, currents, and sea ice that constantly shape and reshape the coastline. Under natural conditions—that is, without the intervention of human activity—the sands exist in dynamic equilibrium, with the processes of erosion and deposition balancing each other. The movement of sand, however, determines the distribution of habitats and species along this gentle coast.

The coastline harbors a triumvirate of productive marine habitats: salt marshes, eelgrass beds, and oyster beds, the last sequestered behind sand dunes in Caraquet, Miramichi, and Bouctouche Bays. Adding to the productivity is the output of the major river systems that flow into the gulf. These natural assets combine to make the region an important spawning, feeding, and nursery area for a diversity of fish species, including Atlantic cod, herring, and mackerel.

## SALT MARSH BUTTERFLIES

THE MARITIME ringlet is one of only two butterflies in Canada that live exclusively in salt marsh habitat, the other being the salt marsh copper. The Maritime ringlet is restricted to fewer than a dozen breeding sites around Chaleur Bay, in northern New Brunswick, and along the southern coast of the Gaspé Peninsula, where the adults fly for a three-week period between late July and mid-August. The small orange butterfly with an eye spot on the underside of the front wing feeds primarily on sea-lavender, though it also exploits seaside goldenrod, sea milkwort, and silverweed as sources of nectar. The caterpillars are green with longitudinal yellow stripes, perfect camouflage for their host plant, salt marsh hay. They pupate at the base of the grass stems and overwinter in the litter layer. The consistent snow and ice cover in the gulf in winter provide insulation against temperature extremes, aiding larval survival, and may be a reason the species has not spread to the Bay of Fundy salt marshes, where tidal range is greater, summer temperatures are lower, and the seawater does not freeze in winter.

MARITIME RINGLET

The lagoonal marsh system that exists behind the barrier islands, in such places as Kouchibouguac National Park, is similar to those found along the coast from Cape Cod to Florida but has no equivalent along the rocky coast of northern New England or elsewhere in the Maritime Provinces. This unique northern habitat was precisely described by the great New Brunswick naturalist William F. Ganong in a 1906 article in the *Bulletin of the Natural History Society of New Brunswick*: "It is across these coves, converting them into lagoons, that the sand islands extend, festooned from headland to headland in great curves inbowed by the power of waves and wind . . . This lagoon is varyingly shallow, with a bottom of sandy mud which supports a great growth of the salt-water eel grass, *Zostera marina*, whose extreme abundance is a characteristic of these lagoons, which is only wanting where salt marshes form in coves or angles, or where river channels wind their sinuous ways to the gullies."

In contrast to the macro-tidal environment of the Bay of Fundy, with its large salt marshes, the Northumberland Shore and other eastern areas of the Magdalen Shallows, as micro-tidal environments, have smaller marshes but support large sea meadows of eelgrass that grow in a number of the coastal lagoons along the eastern New Brunswick shore and on both sides of the

The rich mosaic of habitats, including salt marsh and lagoon, accounts for the southern gulf being "a place of great abundance."

Northumberland Strait. The eelgrass provides a shimmering green glitter to waters in summer and piles up in impressive black windrows in the fall, which some Maritimers still harvest to bank their houses against winter's blasts.

Eelgrass is a green flowering plant that has "returned to the sea." Not a true grass, it nevertheless has ribbonlike leaves that enclose a stem at their base. The blades contain a series of air canals that transport oxygen throughout the plant, including the roots, which grow in soft sediment that is often starved for oxygen. The air canals also confer buoyancy to the plant, thereby orienting the leaves toward the light. Eelgrass grows from an extensive system of rhizomes, or creeping runners, which also put down fibrous roots that anchor the plants. It reproduces by unique flowers that have adapted to function underwater. In late spring or early summer, eelgrass puts out small flowers in spikelike clusters that house both pollen-bearing male and seed-bearing female flowers. However, self-pollination is prevented because the male flowers only release their sticky threads of pollen after the female parts have already matured. The clusters detach and float to the surface where they are transported by the currents. Ripe fruits are shed during this time, and the seeds sink to the bottom. Some of the seeds released from the ripe fruit germinate in early autumn, but most will do so the following spring, avoiding freezing or, in the gulf, the grinding action of ice.

Eelgrass is a perennial plant, and in some cases it is more productive than salt marsh cordgrass. Few herbivores consume the living blades directly, the exception being Canada geese and brant, which gather in large flocks to feed on the eelgrass in the fall, fueling their southward migrations. Most of the productivity from eelgrass enters the detritus food cycle. Eelgrass is constantly shedding old blades, which are broken down by bacteria. The decayed organic matter, or detritus, is then ingested by a variety of detritus-feeding snails, bivalves, and amphipods. They pass the refractory detritus through their digestive tracks virtually unchanged, but in the process strip from it the nourishing bacteria, protozoa, and microalgae. Their fecal pellets are then colonized by bacteria and are themselves recycled.

Eelgrass beds perform other vital ecological functions, acting as a nursery area. The three-dimensional complexity of eelgrass beds, as compared with unvegetated habitats, increases the habitable space, but more importantly, provides protection from predators. Juveniles of many fish species have been found in the gulf's shallow lagoons: among them, Atlantic tomcod, winter flounder, striped bass, stickleback, cunner, and white hake. Sand shrimp and rock crab juveniles also find shelter in the eelgrass beds. While reducing the foraging efficiency of predators, the high productivity of eelgrass lagoons also attracts a variety of birds that prey on small fishes, and consequently, double-crested cormorants, blue herons, gulls, and terns can often be seen hunting in the shallows.

Among the organisms that find food and shelter in the gulf's eelgrass beds are sticklebacks (left) and green crabs (right), the latter an invasive species that has caused widespread damage to eelgrass.

In recent years, European green crabs have been invading northern eelgrass beds, which they also utilize as a nursery. Molecular evidence indicates that these crabs originated from the cold waters of northern Europe and probably arrived in the southern gulf in ships' ballast. The highly adaptable green crab is a voracious predator and has had a major impact on native species. Declines in eelgrass beds in the gulf region have been strongly associated with abundant and increasing numbers of green crabs, which disturb the sediment and thereby uproot eelgrass plants.

Between 1930 and 1933, a wasting disease decimated eelgrass beds on both sides of the Atlantic, causing the loss of 90 percent of beds along the eastern seaboard. The culprit proved to be a pathogenic slime mold, although other stress factors were probably involved. The loss of eelgrass caused widespread erosion, but it had its most devastating impact on brant, which relied heavily on eelgrass as a food source. Their numbers nose-dived, and to survive they switched their diet to widgeon grass and sea lettuce. Many of the eelgrass beds have recovered, having been seeded by plants that survived in the fresher parts of the estuaries, where the pathogen could not tolerate the low-salinity environment.

DENYS ACCURATELY DESCRIBED the topography of the coast as "great flats of sand which advance far out into the sea." One of the places he explored was the estuary of the mighty Miramichi River. Here again, he gives us an insight into the great fertility of these waters: "If the Pigeons [Passenger pigeons] plagued us by their abundance, the salmon gave us even more trouble. So large a quantity of them enters into this river that at night one is unable to sleep, so great is the noise they make falling upon the water after having thrown or darted themselves into the air."

Although Atlantic salmon numbers have been drastically reduced since Denys's time—not only on the Miramichi but throughout the world—by a variety of human-induced factors, the Miramichi remains the greatest producer of Atlantic salmon in North America. In recent years, between 40,000 and 160,000 salmon have returned to its waters after spending one to two years feeding at sea. As recently as the 1930s, that number was close to a million spawning salmon. Great damage was done to the stocks by the budworm spraying programs, which began in the 1950s with the drenching of New Brunswick's affected forests with DDT and continued until the 1990s with various other organochlorine pesticides.

The four main branches of the Miramichi system, along with some twenty-six tributaries, drain one-third of New Brunswick and provide 1,000 square kilometers (386 square miles) of spawning habitat. This vast watershed is sparsely populated and has not been dammed, as many other salmon rivers have been in the species' traditional range from Connecticut to Labrador. Even so, the system produces far fewer salmon than the optimum number—estimated at between 300,000 and 400,000—because of limiting factors both on land and in the oceans, including poor forestry practices and possibly climate change.

In boreal estuaries like the Miramichi, the smelts and salmons are the dominant fish families. Along with the sturgeons, basses and sea lampreys, they have evolved the anadromous habit of moving from salt water to freshwater to spawn. Anadromy is the rule rather than the exception among estuarine fishes in northern waters, but this behavior declines in tropical waters. This trend may relate to the principle of competitive exclusion, in which interactions with other fauna are more important than the physiological adaptability of individual species in making the transition from salt to freshwaters for spawning. At high latitudes there are fewer freshwater fish species than in the tropics. Two factors may, therefore, discourage anadromy in tropical waters: predation pressure and the increased demand for spawning space.

In northern latitudes, anadromous species utilize the brackish estuarine waters in different ways, depending on the phase of their life cycle. Adults use the estuary as a resting place where they can acclimate themselves to the freshwater before making their spawning run. The estuary performs an even more critical function for the young. Shads, smelts, and bass quickly drop back into estuarine waters, where the rich production of small life-forms yields ample food for growth. The young of Atlantic salmon may spend several years in their streams and rivers of birth before heading out to sea, but the estuary may serve as an important way station for feeding.

In the Gulf of St. Lawrence, anadromous species also follow a predictable spawning regimen. Gaspereau (or alewives) make their upriver spawning runs in May and June, and salmon surge into most rivers bordering the gulf in October and November. Some gulf rivers, like the Restigouche, Miramichi, and Margaree, also have a spring run of spawning salmon that enter the river in June. The Miramichi also has a spawning population of striped bass, which enter the river in late May to the middle of June. Finally, smelt migrate into estuaries and bays in late fall and overwinter there until March.

A series of barrier beaches and islands have developed along the shoreline in the southern Gulf of St. Lawrence. Because of the semi-enclosed nature of

## GOING TO SCHOOL ON SMELTS

SMELT OCCUR on the Atlantic coast from New Jersey in the south to Hamilton Inlet, Labrador, in the north. They are most abundant in the Canadian Maritime Provinces, however, in the middle of their range. As a child I helped my family net smelts on their spawning runs, which was considered a rite of spring. Most studies of their biology and distribution were done before the 1940s, when they were considered a more important food fish. More recently, though, the large spawning population of the Miramichi has been the subject of intensive scientific research, providing insight into the anadromous lifestyle.

American, or rainbow, smelt belong to the fish family that includes the eulachon of the west coast and the capelin in east coast waters. Ranging in length from 15 to 20 centimeters (6 to 8 inches), this slender fish has a dark green or olive-green back and rainbow-colored sides—iridescent purple, blue, or pink. The lower jaw projects beyond the upper one, and the tip of the tongue sports large teeth. As adults, smelt eat shrimplike organisms, aquatic worms, and small fishes, including juvenile herring, mummichog, and silversides.

Smelt are a schooling species and, sensitive to light and warm temperatures, tend to congregate near the ocean bottom during the daylight hours. They leave the estuaries en masse to make their spawning runs as soon as the ice leaves the brooks and creeks in spring. The spawning run lasts only a few weeks. Many males die shortly after spawning, and the surviving males and females stay on the spawning ground for five to ten days before moving downstream again. In the Miramichi system, they remain in the bay for the summer but begin moving into the estuary in October, taking up winter residence in the lower estuary after the ice forms in late November.

Smelt spawn at night and demonstrate a homing behavior, tending to return to the same stream, or one nearby, year to year. They lay their eggs on the gravelly bottom, females depositing from seven thousand to seventy thousand eggs, depending upon their size and maturity. Two males clasp the female between them during spawning, releasing their milt simultaneously as the female releases its eggs.

The outer envelope of the egg becomes sticky on contact with the water and adheres to the gravel. When this outer coat is torn away by the current, a portion of the egg, a little stem, sticks to the bottom, allowing the egg to sway in the current, where it is aerated. In some cases, smelts may deposit thick carpets of eggs, and those on the bottom are smothered, leading to high mortalities. Even when eggs receive adequate oxygen, only about 4 percent hatch. The time required for hatching is dependent on water temperature: at 4°C (39°F) eggs require at least

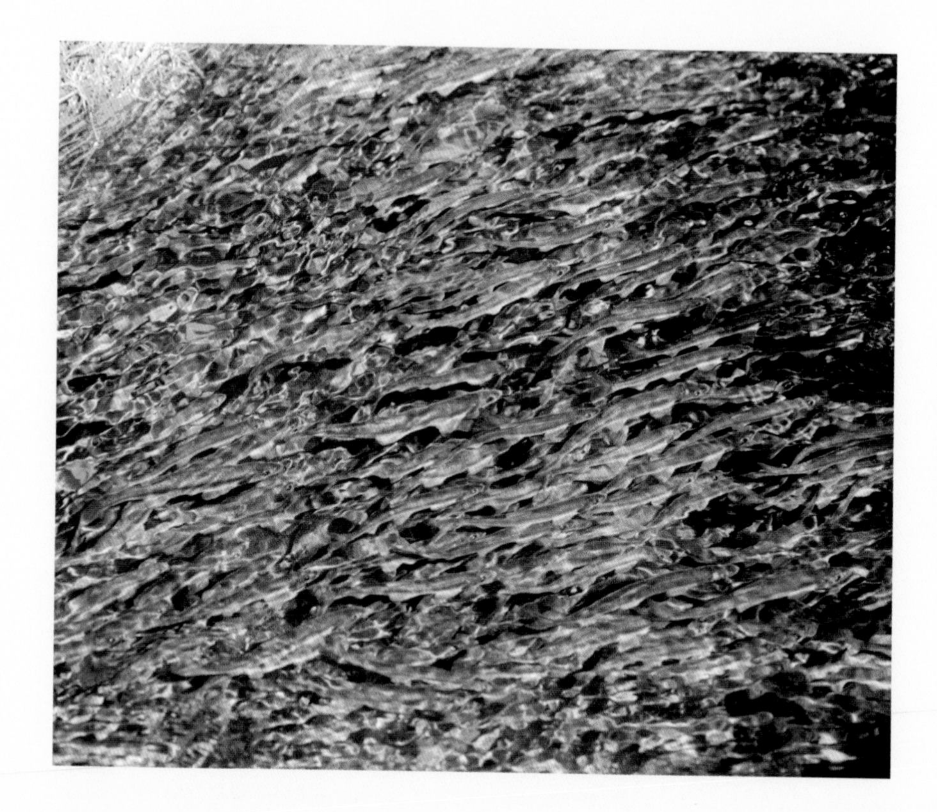

A living river of smelts makes its way upstream from the ocean to spawn in freshwater.

fifty days to develop, whereas at 10°C (50°F) they may hatch in as little as twenty days.

The 5-millimeter-long (0.2-inch) hatchlings are carried downstream to the estuary, where they drift on the tide. Larvae and juveniles are found throughout the Miramichi River, bay, and estuary during the summer, but begin leaving the river in August. The smelt larvae consume small zooplankton and are prey to larger fish. Adults are eaten by a wide variety of predators—cod, salmon, seals, cormorants, and mergansers—and are an important element of the estuarine food web.

the gulf, the shoreline is relatively protected from the wave action of the open Atlantic. There is less wear and tear on the shoreline because the southern gulf is also ice-covered for at least three months of the year, beginning in late December. Furthermore, the prevailing winds are westerly, and so the fetch of the wind is short along most of the shoreline, with the exception of the Magdalen Islands, which are more exposed. These features differentiate the barrier island and dune systems of the gulf from those systems that prevail along the eastern seaboard, from Cape Cod to Florida, which is fully exposed to the open Atlantic. Otherwise, the systems have much in common.

*facing page:* An American bittern strikes a pose to hide in marsh grass but his "song" is a giveaway, earning it the nickname "stake driver."

The centerpiece of this dramatic coastline is the Prince Edward Island National Park, a narrow coastal strip of 32 kilometers (20 miles) protecting a diversity of coastal habitats: in turn, beaches, sand dunes, brackish ponds (or barachois), and maritime coastal forests, as one progresses from the tide line toward the land. These seemingly different habitats are produced by a single process, which begins with the erosion of the soft sandstone sediments that make up the island's red coastal cliffs and soils. Wave action from the gulf is constantly eating away at this fragile bedrock, breaking it down into sand particles. This sets in motion a dynamic process whereby the sand is first transported offshore and then deposited at the surfline to form the barrier islands, beaches, and dunes that string along the Island's North Shore.

Wind and water constantly reshape this coastline. The water is very shallow, only 16 meters (52 feet) deep as far offshore as 16 kilometers (10 miles), and for this reason, the wind-driven waves easily re-suspend the sand and silt and carry it toward the shore, where it piles up above sea level to form offshore bars and sandspits. These barrier beaches sometimes form offshore islands such as Hog Island, at the mouth of Malpeque Bay. However, storms, abetted by rising sea levels, have acted together to move much of the barrier island system, depositing it as landfast barrier beaches. Once the sand is above the waterline, onshore winds whip up the sand particles and carry them farther inland, the first step in the creation of a dune.

As a loose assemblage of sand particles, a dune is inherently unstable. It is the adaptations of sand-loving marram grass that stabilizes the dune. While the leaves and stems act as sand catchers, the rhizomes branch out to form a root system several meters in length, which anchors the dune, creating a relatively stable landform. The foredunes are higher at their leading, windward edge and taper on the landward side, in the lee of the wind, forming a crescent shape. The dunes run parallel to the shore and are subject to constant sculpting and reshaping. Exposed to the forceful onshore winds, the face of

The roots of marram grass knit together the dune sands, which are vulnerable to wind and wave.

the dune is bare sand, but the lee side supports luxuriant growths of marram grass. Storm winds can cause "blow-outs," however, undermining the marram root system. And as the face of the foredune grows higher, the supply of sand on the lee side is diminished, slowing the growth of the marram.

Farther away from the shore, a variety of salt-tolerant plants begin to take root on the leeward side of the dune and on the secondary dunes that often form behind the primary one. Among them are seaside goldenrod, common wild rose, and beach heather, which add splashes of color to the seaside scene. Eventually, soil begins to form on the lee side of dune systems, resulting in a transition to a more terrestrial flora, with the growth of shrubs such as northern bayberry. This shrub bears buckshot-sized fruit covered in a waxy coating favored by birds such as the myrtle (or yellow-rumped) warbler, which takes its name from the plant, sometimes known as myrtle. Both bayberry and beach pea are capable of nitrogen fixation, the conversion of atmospheric nitrogen into organic nitrogen compounds. They achieve this through a symbiotic relationship with nitrogen-fixing bacteria. This nitrogen is key in the nutrient sustainability of the dunes.

In rare instances, a parabolic, or wandering, dune system develops. One such system is the Greenwich Dune, now an adjunct to the Prince Edward Island National Park. This dune is moving from the tip of the Greenwich peninsula in the direction of the maritime forest behind it. The leading edge of the dune is traveling at the dreamlike, but inexorable, rate of 3 to 6 meters (10 to 20 feet) per year. Evidence of its travels is seen in the skeleton forest left in its wake. The tops of white birches stick out of the sand like candelabra gracing an immense white tablecloth. Characteristically, a parabolic dune is horseshoe shaped, with the prongs of the U pointing into the wind. The inside of the U is hollowed out as wind picks up sand grains and moves them to the crest, depositing them on the far side. In this manner, the dune moves, or is said to wander, while maintaining its shape. The lee side of the Greenwich dune contains a natural plant succession, from marram to shrub growth such as bayberry, and, finally, to wind-stunted white spruce. Three large barachois ponds are also located behind the wandering dunes. The brackish waters of the ponds attract a variety of waterfowl and other marsh birds such as American bittern and pied-billed grebe. Savannah sparrows, short-eared owls, and horned larks also haunt the dunes.

*Warm and Cold*

Living near the Northumberland Strait, for four months of the year I look onto an Arctic-like icefield, blindingly white against an azure or pewter sky. In certain years, pans of ice may persist into May. The Gulf of St. Lawrence differs from other large estuaries along the North Atlantic coastline in that it is ice-covered in winter. This ice has three principal sources: locally formed icefields; ice from the St. Lawrence River and estuary; and ice, including small icebergs, that move from the Labrador Shelf into the gulf via the Strait of Belle Isle. Ice begins to form in sheltered areas in December and increases rapidly throughout January. By the end of January, ice covers about half of the gulf, and the entire gulf is blanketed in pack ice by the end of February. The ice breaks up rapidly in the spring as the prevailing westerly winds and water currents carry it southeastward out of the gulf toward the Scotian Shelf.

This ice plays an important ecological role, as it becomes a breeding platform for three species of seals: harp, gray, and hooded. Of these harp seals are the most numerous. They belong to the group of hair seals, family Phocidae, which are characterized by negligible external ears, hind flippers which do not

A harp seal suckles its pup on the birthing ground of a Gulf of St. Lawrence ice floe.

turn inward, and the absence of underfur—precisely the three features that distinguish them from the eared seals, Otariidae. In addition, harp seals are distinguished by having white-coated young.

Harp seals are cold-water animals well-suited to the North Atlantic, where there are three stocks: Newfoundland, Jan Mayen Island, and the White Sea. The Newfoundland stock is further divided into two substocks, one which whelps in the Gulf of St. Lawrence and the other at the so-called Front, east of Labrador and northeast of Newfoundland. These stocks are not separate genetically, and although they intermingle, they maintain a Front-to-Gulf ratio of 2:1, probably related to relative amount of whelping habitat. The greatest year-to-year changes in this ratio are due to fluctuating ice coverage in the gulf, which may be thick or almost absent, depending upon climatic conditions.

In late autumn, mature seals migrate from their Arctic summering grounds, near Baffin Island and the west coast of Greenland, to feed primarily

Île d'Entrée, one of the Magdalen group, greeted Jacques Cartier on his 16th-century voyage of discovery when he marveled at the gulf's natural wonders.

on capelin in the gulf for two to three months before the breeding season. Animals that breed near the north shore of Prince Edward Island, for example, may migrate 3,000 kilometers (1,860 miles) during their annual cycle. It has been shown that all animals that winter in the gulf were born in the gulf. However, many gulf-born immatures—one-to-three-year-olds—winter on the Front, where there are higher biomasses of capelin. In summer, though, the prime feeding area for young-of-the-year harp seals is in the estuary, where they also feed on capelin. The only area where harp seals feed predominantly on herring is near the Magdalen Islands.

Gray seals also exploit the icefields in the gulf for breeding, though not in as great numbers as harp seals. They give birth on the newly formed ice in the eastern part of the Northumberland Strait and along the western shore of Cape Breton Island. Small whelping concentrations are also found on Amet Island in the Northumberland Strait and on Deadman Island in the Magdalen Island group.

HOODED SEAL

Pups are born on the ice from late December to February, in widely scattered groups. The females aggressively defend their young, which may fall prey to bald eagles and ravens if the birthing grounds are close to the mainland. At the end of the fifteen- to twenty-day nursing period females become receptive to mating; males then leave the colony, except for a few old, dominant animals. When the gale-force winds that whip the gulf in winter cause the break-up of the ice, females drift with their young out of the gulf. Before returning to their breeding and whelping areas, to which they show a high degree of fidelity, the seals of all ages undertake extensive wanderings around, and in some cases outside, the gulf.

Hooded seals also occur in much smaller breeding units than harp seals. Hooded seals are larger than both harp and gray seals, growing to 2.7 meters (9 feet) and 400 kilograms (900 pounds). Their most distinguishing feature is the large nasal apparatus of the males. Part of the nasal cavity is enlarged to form a hood, which runs from the crown to the upper lip and overhangs it in older animals. When angered, the seals can enlarge this cavity, and they also can inflate the nasal membrane, which appears as a red balloon-like sac protruding from the nostril. They congregate near the Magdalen Islands, where they have a one-month breeding season, spending most of their time on the pack ice, before migrating toward Greenland in early May. It is unclear whether they feed while in the gulf, but if they do, they probably exploit deep-water species such turbot, cod, and redfish.

IN SUMMER, THE Magdalen Shallows become a warm-water refuge. During the warming period that followed glaciation, warm water extended northwards along the coast as far as the Gulf of St. Lawrence, and with it species more characteristic of the Virginian province to the south of Cape Cod. One of the species that thrived north of its current distribution was the American oyster. Today it survives in the Maritime Provinces only in shallow water areas of the Gulf of St. Lawrence, with the exception of a pocket in the Bras d'Or Lakes. Extinct populations—known from fossil shells in the Minas Basin, on Sable Island, and in the Magdalens—speak to its formerly more widespread distribution, and large pre-colonial middens along the Maine coast confirm its former abundance there. Oysters can tolerate low winter temperatures, as is the case in the gulf, but require warm summer temperatures, up to 32°C (90°F), to breed and propagate. Because the southern gulf is shallow, the waters are warm enough in summer to support oysters. They can feed and grow at temperatures below what it requires to produce eggs and sperm and therefore may only produce spawn during the summer. In northern waters, below 4° to 5°C (40°F), they lie dormant.

The oyster's reproductive behavior is one of the most remarkable in the animal kingdom. It spawns first as a male and later switches to being female and produces eggs. It may make this sex change during successive seasons. As a female, it may produce as many as 100 million eggs. The oyster larvae, known as "spat," settle to the bottom, where they attach themselves to any hard surface, often other oyster shells, which in time produce an oyster bed. The oyster has a byssus gland on its foot that discharges a sticky secretion, cementing it to the substrate. It always lies on its left shell, which is more bowl-shaped, and thus keeps its edge raised above the bottom, preventing it from being fouled with silt.

IN JUNE 1833, almost three centuries to the year after Cartier first reconnoitered the remote, flat-topped Bird Islands in the center of the gulf, John James Audubon recorded his approach to the islands, which at first appeared to him "to be covered with snow several feet deep." It was, however, a flurry of gannets. "I rubbed my eyes, took up my glass," wrote Audubon, "and saw that the strange dimness of the air was caused by innumerable birds, whose white bodies and black-tipped pinions produced a blended tint of light gray. When we advanced within a half mile, this magnificent veil of floating gannets, plainly seen, now shot upwards into the sky, then descended as if to join the feathered masses below, and again veered off to either side and swept over the surface of the ocean." Today, the gulf remains a place of abundance and wonder.

# 7

# GREAT CURRENTS AND GRAND BANKS

## *Newfoundland and Labrador*

NATURE YIELDS ECSTATIC moments. Several years ago I visited St. Vincent's Beach, on the Avalon Peninsula in southeastern Newfoundland. Here, the crescent cobble beach drops abruptly into the deep waters of the northwest Atlantic instead of shoaling gradually, as is most often the case. The cold waters were not only unnaturally deep but clear, and I hardly could believe my eyes as I gazed through this natural aquarium window and watched as several humpback whales, flashing their great white flippers and pleated throats, rushed headlong toward the shore, within a few meters of the water's edge, then veered away at the last moment. They lunged and spouted as they made foray after foray, each time gulping down great hoppers of feed.

I knew their prey was a little fish, the capelin, waiting for its chance to come ashore and perform its most unfishlike spawning ritual on the beach itself. The capelin was also the reason for the great rafts of seabirds near the shore as I made my circumnavigation of the peninsula that day. Here in the waters of Newfoundland was the perfect convergence of oceanographic conditions to create this natural spectacle—one that is repeated annually in the great turning of the seasons.

Cold Arctic waters from Hudson Bay and the Davis Strait converge off Cape Chidley at the northern tip of Labrador to

< Greenland-born icebergs drift through the northern waters of Labrador and Newfoundland for much of the year.

A humpback whale breaches, a spectacular but mysterious behavior characteristic of the species.

*facing page:*
MAP: Newfoundland and Labrador, including fishing banks and major seabird colonies.

form the Labrador Current, which flows southward along the Labrador and northeastern Newfoundland coasts. At the southern edge of the Grand Banks, 330 kilometers (200 miles) off Cape Race, this icy river in the sea meets its antithesis, the great northward flowing Gulf Stream. Both of these titanic forces—the one Arctic, the other tropical in origin—are deflected, but in their meeting they set the table for a great marine banquet.

The Labrador Current itself consists of a warmer and saltier offshore branch, with origins along eastern Greenland, and a colder and fresher inshore branch, with its genesis in Hudson Bay and the Canadian High Arctic. In its journey from the Arctic, the Labrador Current picks up freshwater from the outflow of the Hudson Strait, which is largely driven by rivers draining into Hudson and Ungava Bays, then collects more freshwater from rivers draining into coastal Labrador. The accumulation of all of this river water makes the Labrador Current one of the major contributors of freshwater to the world's oceans, exceeding by as much as ten times the volume of freshwater flushed into the Gulf of St. Lawrence. Even so, this freshwater component is less than 1 percent of the total volume of the current as a whole.

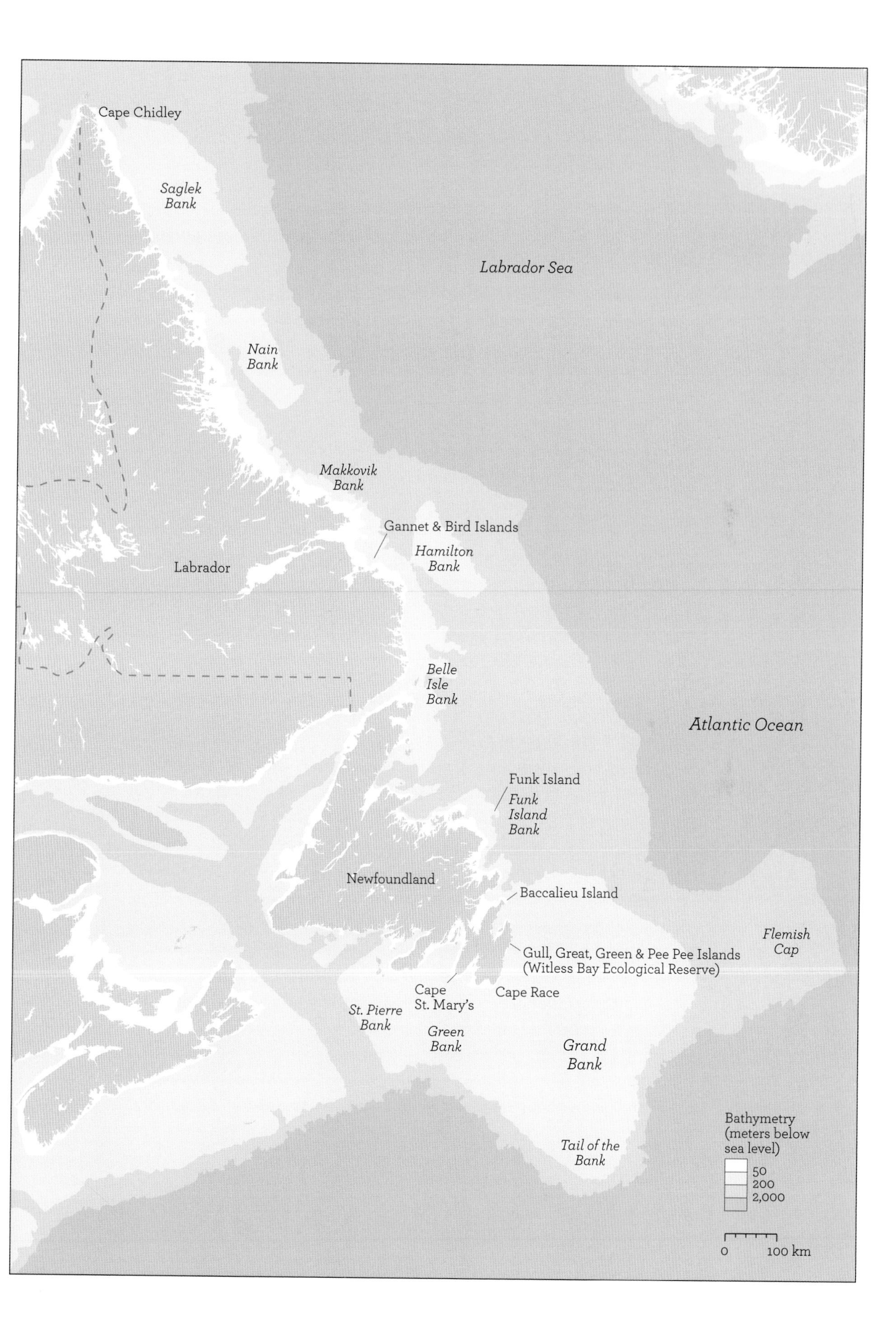

Cape Chidley
Saglek Bank
Labrador Sea
Nain Bank
Makkovik Bank
Gannet & Bird Islands
Hamilton Bank
Labrador
Belle Isle Bank
Atlantic Ocean
Funk Island
Funk Island Bank
Newfoundland
Baccalieu Island
Flemish Cap
Gull, Great, Green & Pee Pee Islands
(Witless Bay Ecological Reserve)
Cape St. Mary's
Cape Race
St. Pierre Bank
Green Bank
Grand Bank
Tail of the Bank
Bathymetry (meters below sea level)
50
200
2,000
0
100 km

The offshore branch is deflected northward, where it becomes part of a counterclockwise gyre in the Labrador Sea. The inshore branch hugs the coast of Newfoundland, before an offshoot turns the corner into the Gulf of St. Lawrence, skirting the south and west coasts of the island and transporting cold waters to the North Shore of the Gulf of St. Lawrence. The end result is that the island of Newfoundland is wrapped in Arctic waters, which, at the cold heart of the current, are just above the freezing point of seawater, at close to -2°C (28°F). The major portion of the Labrador Current then continues on its southward journey, flowing over the Scotian Shelf before spending itself somewhere near Cape Cod. Meanwhile, the Gulf Stream turns to the northeast as the renamed North Atlantic Current, which will eventually warm the northern countries of Europe.

But where these two great currents meet and mingle off Cape Race on the southeastern tip of Newfoundland, a great profusion of marine life is kick-started on the Grand Banks. The Labrador Current is rich in nutrients and carries the plankton that are the base of the banks' grandness. The waters of southeastern Newfoundland have four to five times the biomass of zooplankton of the warmer waters of the Gulf Stream, which nevertheless adds heat, boosting their productive capacity. The productivity of these Newfoundland waters is spectacularly expressed not only in fish but also in the seabirds and marine mammals that gather here in phenomenal numbers.

### *"Fish" by No Other Name*

The continental shelf off southeastern Newfoundland is one of the largest in the world—a vast apron larger than the island of Newfoundland itself. This is the legendary Grand Banks, where the two major currents of the northwest Atlantic converge, stirring nutrients into the water column. The shallowness of the banks, which range in depth from 25 to 100 meters (80 to 300 feet), combined with their great extent, underlies one of the most productive ecosystems on the planet. As fisheries scientist George A. Rose has written in *Cod: The Ecological History of the North Atlantic Fisheries*, "these ecosystems produced an abundance of life the likes of which the world has seldom seen." The great quantities of cod were the draw for most early European excursions to North America. The Grand Banks' capacity as a natural fish-producing factory was probably unmatched in history, as illustrated by the account of John Cabot, who, in 1497, reported the waters "swarming with fish, which can be taken not only with the net, but in baskets let down with a stone."

In the Newfoundland idiom, "fish" has only one meaning, codfish, suggesting the historic importance of cod to the culture of Newfoundland and the central role it has played in the ecosystem of the northwest Atlantic.

Atlantic cod is a generalist and has adapted to many niches in the North Atlantic, where it is a dominant predator on both sides of the northern ocean. Cod are survivors. Part of their durability as a species is their omnivorous appetite. They will eat almost anything, including smaller cod. But their preferred prey is the smeltlike capelin, and wherever capelin have gone, cod have followed. Before capelin emigrated from the Pacific, it is likely that cod populations subsisted on herring, which remains an important alternative prey. But the ability of cod to grow into the huge populations depended on the arrival of capelin, which, like herring, is a fat-rich species and therefore an important source of calories in the cold environment of the North Atlantic.

Given cod's economic importance, it is surprising how little was known about its reproductive behavior until quite recently. Cod tend to form large aggregations in the winter, lying near the ocean bottom in a kind of torpor, at incredible densities of more than one fish per cubic meter (1.3 cubic yards) of water. As spring and spawning season approach, they become more active and form new aggregations, with males and females of similar sizes seeking each other out.

The inshore cod fishery sustained coastal outports for half a millennium in Newfoundland.

Recent research has shown that cod courtship is a ritualized affair, not the random process it was once thought to be. Male cod stake out the spawning grounds first and await arrival of the females to form spawning columns, in which males and females ascend and descend together as if performing a courtship dance. Sound may also play an important role in the spawning ritual. Male cod depress their pectoral fins, creating a sound from their swim bladder: the bigger the fish and louder the sound, it appears, the more successful the male suitor is. For whatever battery of reasons, larger females seem to prefer larger males and may even require them for optimum breeding.

An average-sized female (80 centimeters, or 30 inches) will produce 2 million eggs (a large cod of about 130 centimeters, or 50 inches, could produce as many as 20 million). The round, buoyant eggs rise to the surface, where they float until the little cod hatches after a period of days or weeks. The larva survives the first week of life on its attached sac of yolk. Once the yolk sac is absorbed, the larval cod must also fend for itself by finding small zooplankton or other fish larvae (icthyoplankton) while it is still at the mercy of the currents. Not surprisingly, many perish at this vulnerable transition phase. By fall, the survivors, which may now have drifted far from the spawning grounds, abandon the pelagic world and seek shelter on the seafloor, often in coastal areas.

Rocky substrate or eelgrass beds provide cover from potential predators. In their second year, juveniles venture out to feed, becoming more vulnerable to predation from seabirds and from other fishes, including larger cod. By the end of their second year, they leave their nursery areas and begin to form even-aged aggregations known as shoals, which will be their habit throughout their lifetimes. By the third year they begin to follow the great spawning and feeding migrations of their tribe and by six years old they enter the spawning population. Although only one cod hatched from the million eggs spawned reaches reproductive age itself, it is enough to sustain the population.

Opportunists that they are, cod have adapted to all available spawning habitats, offshore on the continental shelf and in the bays of coastal waters. They have occupied all of the offshore banks, from northern Labrador to the Grand Banks, to the banks of the Scotian Shelf and Georges Bank. Each stock shows a gradual adaptation to cold waters, north to south—a so-called genetic cline—evolving either differing levels of antifreeze in the blood or different migratory patterns. What these populations have in common is a preference for relatively warm water on the banks (2° to 4°C, or 36° to 39°F) for spawning. By contrast, most cod stocks in the eastern North Atlantic spawn in inshore

waters, which are bathed with the warm waters of the North Atlantic Current (as the Gulf Stream is called after it passes the Grand Banks). The temperature of the Labrador Current only partially accounts for the cod's preference for the banks on this side of the Atlantic. The spawning sites also evolved due to the direction and force of the currents, which act like a conveyor belt, delivering the juveniles to the proper place in the ecosystem, at the right time, for their development.

The dramatic coastline of Newfoundland has earned it the affectionate sobriquet, "The Rock."

Smaller coastal populations, however, have developed in the northeast coast bays of Newfoundland—Notre Dame, Trinity, and Bonavista—as well as Placentia Bay in the south. Fish overwinter in these sheltered bays, spawn in the spring, and wait for the spawning populations of capelin to come to them in the summer. Fish on the Grand Banks follow a current gyre, feeding on the offshore populations of capelin and sand lance as they go and in winter only have to swim a short distance to the edge of the banks to escape the cold waters of the Labrador Current. Other great stocks, from the northern banks and Bonavista Corridor, converge on the northeast coast in pursuit of capelin.

The calamitous collapse of the 1990s has reduced the once great stocks to a minuscule fraction of their former selves. The demise of the northern cod—a species of marvelous abundance and fecundity—can be traced to both environmental and human causes. At first, the finger was pointed at environmental factors—in particular, to unusually cold water temperatures, which might cause both high adult mortality and a low survival rate of juveniles. It is true that warmer conditions, on average, produce better recruitment for all of the major northern cod stocks of the North Atlantic—off Norway and Iceland, as well as Newfoundland. Ultimately, however, it was overfishing that delivered the coup de grace to the cod.

The decimation of northern cod began in the 1960s, during a time of high ocean productivity. Fishing fleets from Canada, France, Poland, Spain, Portugal, and the Soviet Union hammered the burgeoning stocks well beyond sustainable limits. To make matters worse, trawlers fished during the winter and the spring, when the fish were gathering in dense schools on their spawning grounds. It was a frenzy from which the usually hardy cod would never fully recover.

Stocks did rebuild to some extent in the 1970s and early 1980s, after the declaration by Canada of a 330-kilometer (200-mile) economic zone, which excluded foreign fleets from much of the Grand Banks. But a decline in productivity in the ecosystem as a whole after the 1960s resulted in lower growth rates in individual fish and, because there were fewer and smaller adults, a reduction in breeding potential. Then, in the late 1980s, the Newfoundland and Labrador waters turned bitterly cold, resulting in fast and furious changes to the ecosystem.

It is a truism: where go the capelin, so go the cod. It was the arrival of the capelin in the North Atlantic that originally allowed the explosion of the cod stocks that greeted John Cabot and subsequently sustained European and North American fleets, as well as the culture of Newfoundland and Labrador, for five centuries. Capelin disappeared from their northern ranges in the early 1990s, their numbers plummeting by 97 percent during a period of the coldest weather ever recorded, which probably affected their plankton food source. Predictably, they began moving south in search of warmer waters and food, and their prime predator, northern cod, followed.

In 1990 and 1991, acoustic surveys detected a huge migratory school of cod, estimated at 500,000 tonnes—perhaps the last great aggregation of the species—in the Bonavista Corridor, a funnel-like migration route off Newfoundland's northeast coast. This motherlode was swimming at depths of

300 meters (1,000 feet) and measured 30 by 20 kilometers (19 by 12 miles)—and probably represented 80 percent of the existing stock. At the front of the pack were the bigger fish, so-called scouts that were leading the search for food. They were likely following the cod's prime prey, capelin, but became prey themselves. Tragically, they swam into the trawls and nets of domestic fishers and foreign fishers. This last great haul of cod spelled the end for what had once seemed an inexhaustible resource.

Although the environment always plays a role in population fluctuations, some scientists reject the hypothesis that attributes the collapse of northern cod to environmental change. Analysis of data going back a century showed that water temperatures were indeed colder throughout the 19th century, up until the 1920s, when the fishery harvest was as large or larger than that of the 1980s. In the more recent past, there were five years (1972, 1973, 1974, 1984, and 1985) in which the water was particularly cold, but there was no consistent relationship between temperature and the recruitment of young fish to the adult stocks. In two of those years, in fact, recruitment was greater than average. According to this view, overfishing was the only cause of the collapse, which began in the north, where trawlers dragged up tens of thousands of fish as they formed dense spawning schools, and worked its way south—a pattern called "fishing down"—until the predictable end came.

Until recently, surveys showed that there was now less than 1 percent of the historical biomass of northern cod. This stock spawned in early spring in selected areas throughout the continental shelf, from Hamilton Bank in the north to the Grand Banks in the south and then undertook spring feeding migrations to inshore waters in pursuit of capelin. Only this offshore habitat is productive enough and large enough to sustain northern cod populations at historical levels.

Historically, a much smaller population of inshore spawning cod inhabited some of the larger bays around the island of Newfoundland, the largest being Placentia Bay, and some inlets of coastal Labrador. These fish spawned inshore and did not make long feeding migrations. They may be descendants of the original cod populations, which predate the arrival of capelin in the northwest Atlantic. These inshore populations may be the only hope of reseeding the larger offshore banks, where populations are so low that recovery could take many decades, if it happens at all. Around the island of Newfoundland there is some movement between bays of these inshore stocks, and in the past they mixed with offshore stocks on the Grand Banks during the summer, when capelin came ashore.

Genetic differences are very small between these southern cod stocks, and DNA analysis indicates that interbreeding has taken place. It is hoped that the inshore stocks might move offshore when the bays, where their populations are growing, become too small to support them, and there is precedent in fisheries ecology for such expansion. In addition, the recovery of capelin stocks on the offshore banks might lure the inshore cod to offshore waters, where they might repopulate the banks. It remains unclear, however, whether these inshore cod can or will expand their range and re-colonize the banks and the continental shelf.

What is known is that the welfare of the capelin is critical to the future of the cod, as it is to all of the predators at the top of the marine food chain: other fishes, such as Arctic char, Atlantic salmon, Greenland halibut, and American plaice, as well as seabirds, whales, and seals. Capelin is a cornerstone species, meaning any food web of the ecosystem is dominated by capelin, with capelin serving as a major food item in the diets of most larger predators and itself acting as a major predator on smaller prey. This dependence on a single central prey species also makes the ecosystem of Labrador and Newfoundland extremely vulnerable to natural perturbations like climate change as well as human activity such as overfishing.

*Love on the Beach*

Capelin, a small, smeltlike fish of silvery green-and-magenta coloration, has evolved the peculiar strategy of spawning on the beaches. The capelin's spawning act is called the capelin scull in Newfoundland, where capelin are said to "roll" on the beach. When they do, outport Newfoundlanders converge on the beaches with hook and line, as well as buckets and nets, to scoop up this strange harvest of fish out of water. It is one of nature's oddest spectacles to see the little fish throwing themselves onshore in a most unfishlike manner to deposit their spawn in the natural incubator of beach gravel.

Male capelin are noticeably larger than females and have an exaggerated anal fin, which propels them onto their terrestrial spawning grounds. They also have a lateral line of spawning ridges, which are actually elongated scales. They look and feel hairy, like a three-day-old beard. (The capelin's Norwegian name, *lodde*, means "hairy.") These ridges are not merely attraction features, like the hooked jaws of spawning male salmon, but have a practical function. During beach spawning, the smaller female is guided ashore by a single male—or often between two males, like an ocean liner between two tugboats. The spawning ridges, as well as the large pectoral and pelvic fins of the male,

## GIANT SQUID

ON OCTOBER 26, 1873, two fishermen—Daniel Squires and Theophilus Piccot—along with Piccot's son—saw something floating in Portugal Cove, Newfoundland. Thinking it was flotsam, they rowed alongside and struck it with a boat hook. To their horror, the mass convulsed into life. Its fearsome eyes flashing, the giant squid rose up from the water and lunged at the 6-meter (20-foot) skiff, fastening its horny "parrot's beak" to the gunnels. Then a stringy, pallid tentacle with a thick, sucker-studded club shot out from the creature's head and coiled like a snake about the stern of the boat. A stouter arm also seized the boat, and the squid began to submerge, intent on taking the tiny craft with it. Piccot's twelve-year-old son had the wit to grab a hatchet and hack off the tentacle and arm, and the giant squid slipped away, darkening the water with jets of ink, but not before yielding proof that the mythical kraken, or giant squid, written about by authors from Aristotle to Jules Verne, actually existed in nature.

Although giant squid roam the world's oceans, Newfoundland appears to be unique in that it is the only place where they have regularly grounded. Squid belong to the cephalopods (literally "head-footed"), a marine family that includes the octopus and cuttlefish. Perhaps the most remarkable organs of the giant squid are their dinner-plate-sized eyes, which are well adapted for seeing in the water by virtue of a movable lens. Whereas the human eye has a point of focus on the retina, the squid have an equatorial band where they are able to focus, making their vision potentially twice as acute as a human's. Sight, more than any other sense, dominates squids' lives—they are thought to communicate with each other by means of bioluminescent chromatophores, which flash and change color from pale gray to brick red. And their ability to focus over a wide range allows them to track prey and predator alike.

The only known predator of the giant squid is the sperm whale, whose stomach contents often contain fragments of giant squid or, on occasion, a whole animal. Eyewitness accounts of titanic clashes between these two pelagic giants were widely circulated among 19th-century whalers.

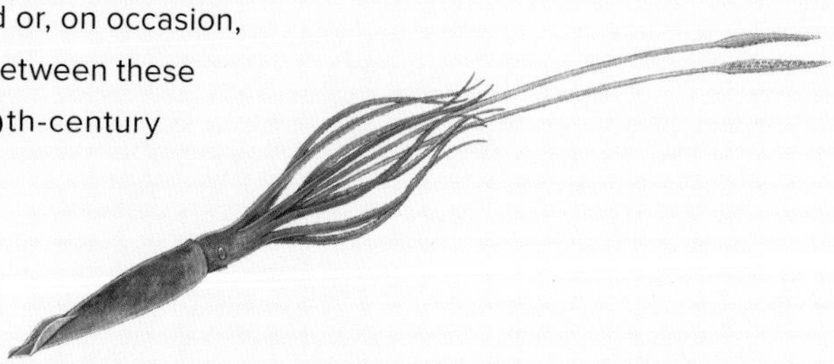

GIANT SQUID

help to hold the female in place for the brief spawning act, which takes place between successive waves upon the beach.

This strange behavior is described in detail in the classic 1948 monograph *The Life History of Capelin (Mallotus villosus O.F. Muller) in Newfoundland Waters*, by Wilfred Templeman:

> When attached the fish head toward the beach swimming vigorously and also taking advantage of the momentum of the wave. They go up the beach as far as they possibly can get in this way and then settle in one spot as the wave recedes, all the time using their fins and tails with great rapidity. In this way they scoop out a slight hollow in the soft sand as if trying to bury the eggs as far as possible, and the vigorous action of the fish in the tiny puddle of water thus formed can be distinctly heard. After separating the capelin lie still on the sand for a second as if exhausted before starting to paddle furiously in an attempt to regain the water, for by this time another wave has rolled up the beach. The writer [G.W. Jeffers] has tried to pick up a pair that was still in contact but found it impossible. Timing the spawning act with a stop-watch shows that it is over in less than five seconds.

*facing page:*
In a most unfishlike manner, capelin come ashore to spawn; their eggs will incubate in the beach gravel.

Most capelin spawn at three to four years of age and then die, though some females may survive. Inshore spawning spans a four- to six-week period during June and July. Most spawning occurs on the beach, but if water temperatures on the beach become too high, some capelin will spawn in the shallows rather than coming ashore. Capelin prefer water temperatures of 5.5° to 8.5°C (42°C to 47°F), and spawning usually occurs at night or on dull, cloudy days. An offshore stock spawns on the southern Grand Banks, underwater. This population is probably a holdover from the last glaciation, when the Grand Banks were emergent and the fish spawned there on the ice-free beaches—and when sea level rose, they continued to use these relict beaches.

Beach spawning seems like a bizarre and risky behavior, but it may be a safer strategy to leave your eggs to develop on the beach than in the water, where predators can get at them. Capelin prefer beaches with fine gravel to which the sticky eggs adhere. The eggs themselves are relatively big, providing the larval fish with a "large box lunch" during their stay in the beach gravel. Depending on the temperature on the beach, the eggs take two to three weeks to hatch, and then the embryonic fish have three to eight days of yolk-sac reserve before they must begin to feed on their own.

The critical event in a capelin larva's life is whether or not an onshore wind blows a warm mass of water into the capelin coves. If so, the rise in temperature triggers its emergence, and the warm water provides an ideal environment for its growth. The water contains a large biomass of small plankton in the edible size range for the still-translucent little fish and contains relatively few predators compared with the deep upwelling water, where the larvae are transported by offshore winds. If no onshore winds occur, the larvae deplete their box lunch of egg sacs and starve; or they emerge into a relative biological desert, where predators such as jellyfish are numerous. More than half of the difference in production from any given year to the next—which can be great, as much as thirty times—can be accounted for by meteorological variations: wind direction, air and water temperature, and hours of sunlight.

Capelin are high in oil and therefore are the main source of fat—read energy—for many marine predators. It would not have been possible for the large populations of cod to develop before the capelin entered the North Atlantic ecosystem. Capelin are also largely responsible for the huge seabird colonies that rim the coast of Newfoundland and Labrador, which boasts among the largest seabird populations in the world. An estimated 10 million seabirds breed here, and some 35 to 45 million use the offshore waters as feeding grounds. Moreover, the distributions of the principal colonies of puffins and other seabirds coincide with those of spawning capelin. In winter, the 4 million thick-billed murres from Greenland and the Canadian Arctic, and many of the estimated 1 million common murres that breed in Newfoundland and Labrador, feed on capelin in offshore waters. Together, breeding and wintering seabirds are thought to consume 250,000 tonnes of capelin each year, the same order of magnitude as consumption by harp seals and fin and minke whales, but only one-tenth of what cod consumed when they were at their peak populations.

### *The Sound and Fury*

All major seabird colonies in Newfoundland and Labrador are found on islands, except the one on Cape St. Mary's, at the southwestern tip of the H-shaped Avalon Peninsula. Its near-vertical 125-meter (400-foot) cliffs provide the nations of seabirds gathered here with ample protection from terrestrial predators. It is the world's southernmost gannetry, the southernmost major breeding site for common murres in the western Atlantic, and the most southerly breeding area for thick-billed murres in the world.

The sound of a seabird colony, and the ammoniac smell, is likely to greet you before the birds themselves come into view, especially at Cape St. Mary's,

which is often shrouded in fog. Seabird colonies are sometimes dubbed "bazaars" because of the cacophony. As you walk over the wind-shorn heath toward the cape, the discordant chorus of seabird cries rises from the cliffs as if out of a natural orchestra pit, ranging from the harsh, high-pitched trisyllabic cries of the black-legged kittiwakes—a small gull that nests on the narrow cliff ledges—to the low, guttural mumbles and murmurings of their neighbors, the murres. Then the birds themselves come into view—first the gannets, great white birds with black-tipped wings shuttling to and from their nesting site atop Bird Rock, a mammoth sea stack split off from the cape. On a clear day, from a distance, the cape looks like a meadow resplendent with a crop of white daisies. At the tip of the cape, you can look across a vertigo-inducing crevice to the top of Bird Rock, where the nesting gannets occupy every square meter. The colony has grown in recent years, so much so that it has overspread to the mainland.

Bird Rock, at Cape St. Mary's, is the most iconic of Newfoundland's many great seabird colonies.

*following spread:* The puffin is the most colorful of the seabirds that breed on islands in the North Atlantic.

The air surrounding the headland is a maelstrom of birds. The gannets describe wide circles out to sea, their streamlined white bodies flashing in

Seabirds, like these black-legged kittiwakes, choose precarious nesting sites that are inaccessible to terrestrial predators.

the sun like the fuselages of distant jets. As you watch, they fold their wings and plunge into the waves, sending up spouts of spray. Larger, bushy exhalations mark the presence of humpback and minke whales as they plow through the waters, gulping down quintals of capelin. All the fishers of the ecosystem are concentrated here, including humans in their boats, which seem dwarfed as they bob in the great sea swells that spend themselves against the foot of the gray cliffs.

Seabirds are perhaps the most adaptable group of creatures on Earth, able to live on land, in the air, and in the water. They have adapted to all the major environments on Earth, from the polar regions to the tropics. They are distinguished from wading birds, like herons and egrets, and from shorebirds, by their ability to thrive in the marine environment, often far from shore and for long periods of time. To do so, they have had to evolve a means of ridding themselves of salt through salt glands, located in the eye orbit, where they act as extra-renal kidneys. In some cases, seabirds spend months, even years, at sea, only coming to shore to breed.

Overall, however, seabirds have several characteristics in common: they breed in vast colonies, are monogamous, and forage over wide distances—all behaviors that have evolved to increase their chances of survival. But seabirds display a number of dramatic differences from passerines, or land birds, in their life history. They are much longer-lived, with life spans of twelve to sixty years compared with five to fifteen for land birds. They begin to breed at a later date (two to nine years compared to one to two), they produce smaller clutch sizes—often one egg only—and their chicks are slow to mature.

Why seabirds are so different in these aspects of their lives is not well understood. It has been hypothesized that seabirds' small clutch size and the slow growth of the chicks may be related to limited food supply near colonies or even, in some cases, to the depletion of food supply by the foraging seabirds themselves. There is little firm evidence to support this "energy-limitation hypothesis," however. The amount of food that adults bring to the nest seems to be regulated by the chick's food-begging behavior. Adults respond to the chick's needs rather than bringing the maximum amount of food to the nest. In addition, both parents are observed to loaf at the nest during the rearing period, suggesting that the amount of food is not the limiting factor. Even if the parents were to bring more food to the chicks, it is not known whether some physiological or genetic constraint would prevent them from assimilating it.

Almost all seabirds—some 95 percent—are colonial nesters, a much higher proportion than in other bird families. (A colony can be loosely defined as a group of individuals at a breeding site.) Gulls, terns, and auks are predominantly colonial. Although colonial living is not exclusive to seabirds, theirs is perhaps the ultimate example of coping in close quarters. Living close together requires constant communication, which is probably why large seabird colonies, sometimes numbering in the millions of pairs, are such noisy affairs, with waves of sound constantly rippling through them.

One obvious advantage of colonial breeding is strength in numbers. Large colonies are better able to defend themselves against predators. For seabirds, however, this advantage probably only applies to avian predators, such as ravens, eagles, or even other seabirds, like the predatory gulls, since they are largely defenseless against mammalian predators, such as cats, dogs, foxes, or even rats. (Extirpation of gannets took place in two southern colonies, on Seal Island off Yarmouth and Gannet Rock in the Bay of Fundy, when domestic pets were introduced in the 19th century by fishers and lighthouse keepers. Inadvertent introduction of rats on the islands of the Atlantic coast of Nova Scotia also decimated the populations of burrow nesting petrels.) The primary defense against mammalian predators is the selection of isolated, inaccessible breeding sites such as offshore islands and steep cliffs.

It has long been thought that one of the effects of colonies is synchrony in the breeding cycles; that is, the breeding activity of other pairs—copulating or courtship—stimulates breeding in nearby pairs. The result is that most chicks are born around the same time, a circumstance that may confer some survival advantages. The chicks also fledge around the same time, further reducing the risk of being picked off by ravens or gulls. The potential predator is said to be "swamped" by the number of potential prey. It is also worth noting that there is a higher survival rate for chicks born near the center of a colony rather than on the periphery, where they are more vulnerable to predation.

Being part of a colony may also help individuals find food, which for seabirds is often far from the nesting site. Birds deprived of food may follow others that they observe successfully bringing food back to the nest or that simply look more fit. Seabirds often have to travel farther for food than other birds—up to 450 kilometers (280 miles) in the case of gannets. Birds seem to follow the successful fishers, just as human fishers use seabirds as signs of where to set their nets or lines. The distances between the nest site and the foraging area explain, in part, why breeding usually requires both the male and the female to be involved in chick rearing—reaching the feeding grounds and collecting food takes more time than it does for land birds. In addition, the need to defend a nest far from the foraging areas probably also led to monogamy, which is predominant among seabirds. For the good of the chicks, someone has to stay home.

Colonial breeding has its drawbacks. It increases competition for food, especially among species such as terns and gulls that forage close to the colony. The crowding also encourages the proliferation of microbes and parasites and the transmission of disease. And crowding can lead to aggression or cuckoldry.

Seabirds are fundamentally different from other marine creatures, such as whales and fish, in that they breed on land. Often their choice of a breeding site is a compromise between an area where oceanographic conditions produce sufficient food and one that provides adequate protection from predators. To improve the efficiency of feeding, they use oceanographic phenomena, such as fronts and upwellings, that concentrate prey. Even so, birds often need to forage within a 35- to 50-kilometer (20- to 30-mile) radius of their colonies. Finding a site well protected from predators, however, is not as easy as might be expected. The southern and eastern coasts of Newfoundland provide some suitable cliff sites, most notably at Cape St. Mary's, as does the rugged coast of Labrador. The largest colonies, however, are restricted to islands off the coast.

*Island Kingdoms*

The largest seabird colony in Newfoundland waters is on Baccalieu Island, just off the northwestern tip of the Avalon Peninsula, near Bay de Verde. It derives its name from the Basque word for codfish, reflecting the productivity of the local waters. The colony hosts eleven species of breeding seabirds, a greater diversity than any other colony in the province, including black-legged kittiwakes, common and thick-billed murres, razorbills, northern fulmars, gannets, and puffins. But its most numerous resident breeder is also its most inconspicuous—the Leach's storm petrel, which numbers more than 3 million pairs, making Baccalieu the largest Leach's storm petrel colony in the world.

A seabird colony is a primal place, all sound and fury, darkness and light.

Petrels are starling-sized, soot-colored birds that forage far out to sea and only return to their breeding islands at night, a strategy evolved to foil potential predators such as gulls. In the past, the sighting of petrels near land—whether Leach's storm petrels or Wilson's storm petrels—was considered a sign of bad weather, and for this reason they were sometimes called "gale birds." The term "petrel" appears to be a corruption of "St. Peter," the apostle who supposedly walked on water, which is what a storm petrel appears to do when it is foraging for zooplankton on the wave tops. A further term applied to petrels in medieval times was Mother Carey's chicks, a sobriquet sailors invoked in the hope that *Mater cara* would protect them from storms at sea.

I well remember my first encounter with the petrels of Baccalieu. I was sitting in the lightkeeper's house on an unusually warm summer night for Newfoundland. The window had been left open in the kitchen to allow the air to circulate. Although I had seen the burrows in the dense stands of balsam fir, where one adult tends the chick while the other is offshore foraging, I had not seen a single adult bird during the day. But shortly after night had fallen, I began to hear the birds returning en masse to the island. Their cries were like the electronic babble of a synthesizer or the anonymous chatter of the rain forest. Then one of the birds accidentally flew in through the open window and was gently captured by the lightkeeper's Golden Retriever, and the lightkeeper released it. Outside, I marveled at the huge flocks of petrels circling the lighthouse like moths to the flame. Unless the moon is out, a petrel cannot locate its burrow—one among millions—by sight, which means that both smell and sound probably play primary roles. Petrels find food by odor, so it is likely they also use smell to locate the colony, then home in on their mate's call to pinpoint their nesting burrow. Once they have located the burrow, they regurgitate the food they have plucked from the waves to sustain the single chick.

After Baccalieu, the largest Newfoundland colonies are on a quartet of islands—Gull, Green, Great, and Pee Pee—in Witless Bay, midway along the eastern, or Irish, shore of the Avalon Peninsula. Protected as part of the Witless Bay Ecological Reserve, the islands provide refuge for the largest concentration of breeding seabirds in eastern North America, including three-quarters of the Atlantic puffins (260,000 pairs) and the largest breeding

Atlantic puffins stand by their burrows in the Witless Bay Ecological Reserve, whose islands are home to the largest concentration of breeding seabirds in eastern North America.

concentration of black-legged kittiwakes in the northwest Atlantic, as well as the second-largest colony of common murres and the second-largest colony of storm petrels (some 620,000 pairs) in the world.

Seabirds have a number of nesting strategies: petrels and puffins dig burrows; gannets build their nests on the ground; murres and kittiwakes are cliff nesters. By occupying these different niches, they avoid competition for space. A trip around the islands by boat—landing on the islands is not permitted except for research purposes—provides an object lesson in how this United Nations of seabirds partitions the limited real estate and cohabits in relative peace. On the brilliantly green slopes of Great Island, legions of puffins stand philosophically by their burrows. The center of the island is partially forested, though many of the trees are dead or dying because of the burrowing of puffins and Leach's storm petrels and the nitrification of the soil from generations of accumulated guano. The other side of the island is precipitous: slabs of rock, like great wheels stacked one on top of the other, pitch into the sea at an acute angle, and high undercut cliffs serve as murre and kittiwake high-rises.

The most conspicuous and numerous seabirds in Atlantic waters off Newfoundland are the auks, or alcids. This family of dapper black-and-white birds includes six living species—common murre, thick-billed murre, razorbill, black guillemot, Atlantic puffin, and dovekie—and the extinct great auk. With some scientific validity, these birds are known as the penguins of the north, since they effectively occupy the same ecological niches as do penguins in the Antarctic. In fact, the term *penguin* was first applied to the extinct great auk, and may have had a number of origins. In Welsh, *penguin* means white-head, which might be descriptive of the white patches in front of the great auk's eyes. Or it might have derived from the Latin *penguis*, meaning fat or grease—the species was rendered for its oil, and sometimes the bird itself was used as fuel. Or it might have referred to its disproportionately small wings through a corruption of "pin-winged" or "pinioned."

At three-quarters of a meter (3 feet) tall, it was by far the largest of the auks and the only one that was flightless. It was the combination of these two traits that probably led to its demise. The great auk's breeding colonies appear to have been quite widespread throughout the North Atlantic, where the birds served as sources of fresh meat, first for trans-Atlantic explorers and later for fishers. It was said that they could be herded aboard ships on gangplanks. Wholesale slaughter of the bird and decimation of its eggs culminated in their extinction. The last sighting of a pair of Great Auks was made in 1844 on Eldey Island, Iceland.

GREAT AUK

Unlike their southern counterparts and the great auk, the other alcids have not lost the ability to fly. The capability for flight and prolonged deep diving appear to be mutually limiting and only coexist in the case of alcids weighing 1 kilogram (2 pounds) or less—which includes all of the surviving auk species. However, both the penguins and the auks (which are unrelated) possess a remarkable proficiency at "flying" underwater, an example of convergent evolution—similar adaptation to a similar environment.

The auk's wing has been modified so that it can be used both as an aerial and a water paddle. It is short, reducing its resistance in the heavy medium of water and optimizing underwater propulsion. All the wing elements have been strengthened, by the enlargement of ligaments and thickening of connective tissue, to adapt to the stresses of underwater diving. In addition, the pectoral muscles are exceptionally large, allowing the bird to compress the breast feathers and thus reduce buoyancy when diving.

The larger the bird, the deeper it can dive. We can only speculate to what depth the great auk was able to dive, but it probably exceeded 100 fathoms. Even its much smaller relatives plumb surprising depths; common murres have been caught in fishing nets at 100 fathoms.

Seabirds have evolved a variety of feeding strategies to exploit all trophic levels of the food web and thus avoid competition. Petrels, like the albatrosses to which they are related, wander the world's oceans, where they seem to dance on water while feeding, plucking zooplankton from the surface—a method called dipping. Gannets soar for hundreds of kilometers in search of their prime fish prey, then fold their great wings, plunge-diving to catch mackerel, herring, or capelin. By contrast, the alcids—murres and puffins—have short, stubby wings that make them poor fliers but excellent divers. They can thus exploit a greater range of depths to find food than other seabird species; this method is called pursuit diving. However, all depend on their prey—capelin in Newfoundland waters—to concentrate near their nesting sites. Some species, like the gulls, are scavengers, feeding on dead animals for at least part of their diet. Still others are aerial pirates, such as the jaegers and skuas, which occur in small numbers in inshore waters, where they are sometimes observed harassing black-legged kittiwakes causing them to disgorge capelin.

Seabirds play a vital role in recycling nutrients in coastal waters. The great Newfoundland ornithologist Leslie M. Tuck has described murre colonies, which can number more than a million individuals, as "the fertilizing factories of the northern seas." Seabird droppings are rich in nitrates and phosphates, which boost phytoplankton growth in the upper 15 fathoms of water, where light penetrates and photosynthesis takes place. This input of fertilizer is particularly important in the summer, when seabirds concentrate at breeding colonies, because the upwelling currents that bring nutrients to the surface are weaker then. In fact, studies show a correlation between fisheries production and the presence of large seabird colonies along the coast of Newfoundland. Seabirds, in effect, help to grow their own food supply as their guano is recycled into the marine food web.

The role of seabird colonies in the early exploration of the North American continent has been compared with that of the buffalo in opening up the West. The most pressing concern for early explorers, having made the two-month-long crossing of the North Atlantic, was the replenishment of fresh meat supplies. Newfoundland's great seabird colonies served as the first North American "fast-food takeouts," according to Memorial University of Newfoundland ornithologist William Montivecchi. The most important of these fast-food outlets was Funk Island, a fact that sealed the fate of the great auk.

The Funks consist of an 800-by-400-meter (2,600-by-1,300-foot) bald island of granite and two rocky nearby outcrops, 60 kilometers (37 miles) north of Cape Freels, Bonavista Bay. They first appear on a 1503 Portuguese chart as *Y.-dos-Aves*, "Isle of Birds." The island's ornithological bounty was first chronicled by Jacques Cartier, who visited the Funks on May 21, 1534, when he found the island still surrounded by ice cakes: "It is so exceeding full of birds that one would think they had been stowed there. In the air and round about it are a hundred times as many more as on the island itself. Some of these birds are as large as geese, being black and white with a beak like a crow's. They are always in the water, not being able to fly in the air, inasmuch as they have only small wings about the size of half one's hand, with which however they move as quickly along the water as the other birds fly through the air. And these birds are so fat that it is marvelous..." He was, of course, describing the great auk. Cartier's crew filled two longboats with the birds in less than half an hour, salting them in four or five casks and eating the rest fresh. It was the beginning of the end for the great auk.

The Funks was probably the largest great auk colony in the world, perhaps numbering more than 100,000 pairs. Cartier also describes the presence of murres and gannets, both of which still nest on the island—some 800,000

Common murres mass in incredible densities on the bare ledges that serve as their nests.

murres and approximately 8,000 gannets. These species usually nest on steep cliffs, but on the remote Funks they nest on flat ground—the only option on this bald, flat, sea-washed rock, which is too remote for mammalian predators. The numbers of each have grown substantially in the last century, having been previously decimated by egging and hunting.

Today, The Funks host 80 percent of the total northwest Atlantic population of common murres, a colony that runs the entire length of the island. The natural history writer Franklin Russell vividly described the density of

the breeding birds when he visited there in the mid-1960s: "The murres were massed so thickly they obscured the ground. The birds stood shoulder to shoulder, eyeball to eyeball. In places, they were so densely packed that if one bird stretched or flapped her wings, she sent a sympathetic spasm rippling away from her on all sides. All life was in constant, riotous motion."

Although murres on The Funks nest on the flat terrain of the island, elsewhere they nest on precipitous cliffs. Their choice of nest site on narrow ledges high above the rocky shore and the sea seems especially precarious because they do not build a nest, but lay their single egg on the bare ground, or more often, on bare rock. Many eggs do roll off their ledges, which happens more frequently in the first few days after they are laid. Murres have been observed to gather a few pebbles together—building a primitive nest—and the pebbles become cemented to the substrate by sediment and excrement, providing some stability to the egg. This practice is more the exception than the rule, however. It appears that the pear shape of the egg makes it more likely to rotate rather than to roll off of the smooth surface of the ledge.

The shape of the egg and the accumulation of sediment and excrement do enhance the egg's stability, but the most important factor is the shift in weight distribution as the egg develops. During incubation, the small end of the egg becomes heavier, causing the large end to rise, which reduces the radius of the curve described by the egg when it is disturbed. The mechanism for this phenomenon is described in Leslie M. Tuck's classic monograph "The Murres": ". . . at the beginning of incubation the egg's center of gravity, because of the small size of the air cell, is close to the large end. Gradually the embryo develops and the size of the air cell increases, causing the movement of the center of gravity toward the narrow end of the egg and a change in the position of the egg on the ledge." Tuck points out that this increase in stability is very important because the possibility of the birds brooding a second egg diminishes dramatically during incubation.

If the egg survives on its precarious perch, the time comes when the chick must take the plunge to the sea. This "sea-going" of the murre chicks is a dramatic and death-defying event, and usually happens during the twilight hours to thwart marauding gulls. Adult birds gather at the base of the cliffs and call excitedly to the chick, which then walks off its ledge, calling shrilly as it plummets downward, furiously beating its still-developing wings. Invariably, it is accompanied by an adult, which seems to hover over the chick with what Tuck describes as a "butterfly-like" flight. Once in the water the chick is mobbed by several adults, jostling and hounding it. Eventually, the crowd disperses and the chick is accompanied out to sea by one or two adults, presumably its parents.

Alcids are also susceptible to a variety of large-scale mortalities from human sources. Some 450,000 murres (or "turres" as they are known to Newfoundlanders) are harvested annually for food in the traditional hunt in the waters off Newfoundland and Labrador. In the past, monofilament drift nets set for both cod and Atlantic salmon exacted a heavy toll on alcids, especially murres and puffins, near their breeding colonies, drowning tens of thousands. The decline in the fishery has been matched by a decrease in this seabird by-catch.

Oil pollution, however, remains a major source of seabird mortality off Newfoundland and especially affects the alcids, which spend their lives far out at sea when not at their breeding colonies. This oil is not from catastrophic spills but from smaller but equally deadly dumping of oily bilge water by ships that follow the great circle route through Newfoundland waters. This illegal but largely uncontrolled activity is equal to an *Exxon Valdez* spill every year—accounting for the deaths of some 300,000 seabirds annually. Sea ducks, such as eiders, and the auks are especially affected. The dumping seems to be more prevalent in winter, when oil slicks are less susceptible to detection because of rough seas and storms. Unfortunately, the winter is also when millions of thick-billed murres that breed largely in Arctic waters gather on the Grand Banks, where they meet their sad fate. The loss of these birds in Newfoundland waters has likely led to the substantial declines recently observed in the Icelandic and Greenland breeding populations.

### *Down on the Labrador*

Straightened out, the Labrador coast would stretch for 20,000 kilometers (12,400 miles). The northern coast—north of Okak Bay—is deeply indented with fiords, and in the far north, the Torngat Mountains rise dramatically and precipitously from the sea to a height of nearly 1,500 meters (4,900 feet), making them the highest mountains in Canada east of the Rockies. This northern region was not covered by the Wisconsinan glacier, but because the whole

coast is composed of Precambrian Shield rocks, which generally resisted glacial erosion, today, most of the coastline is characterized by high relief. Moreover, it is still rising as it rebounds from the relatively recent removal of the glaciers' great weight.

This uplifting of the land—a process known as isostasy—has created a skerry-type coastline, where many bald islands fringe the mainland. The central coast between Okak and Lake Melville is guarded by some four thousand islands, which alone account for one-third of the total shoreline length. The fiords, bays, and passages that thread through this maze of islands extend 60 kilometers (37 miles) seaward in the region of Nain. By contrast, southern Labrador consists of a gentler, more rolling type of landscape where the glacier deposited tills and built up drumlins.

The cold waters of the Labrador Current create a strip of tundra along the outer coast and the islands, which narrows until forest replaces it south of the

The Torngat Mountains—the highest in eastern Canada—rise imperiously from the Labrador Sea.

Icebergs, large and small, drift by the austere Labrador coast for much of the year.

Strait of Belle Isle. In the interior, the forest limit extends farther north, near the latitude of Nain.

For generations Newfoundlanders have gone "down on the Labrador" to fish for cod in the summer. The first European to document life on the coast was George Cartwright, trader and explorer, who lived there from 1770 to 1786. A self-taught naturalist, he provides us with an accurate assessment of the character of the land and sea:

> All the east coast . . . exhibits a most barren and iron-bound appearance; the mountains rise out of the sea, and are composed of a mass of rocks thinly covered in spots with black peat earth, on which grow some stunted spruces, *Empetrum nigrum* [crowberry], and a few other plants, but enough to give them the appearance of fertility; such lands therefore are always denominated Barrens. As some compensation for the poverty of the

> soil, the sea, rivers and lakes abound in fish, fowl and amphibious creatures... All those kinds of fish which are found in the Arctic seas abound on this coast.

The Eurocentric view of a barren land is not borne out by the long history of Aboriginal occupancy. Archaeologist William Fitzhugh of the Smithsonian Institution has written of the Labrador coast that "it seems impossible to find a spot which has not been modified in some tangible way by Inuit hands." Contemporary Inuit have continued to exploit the land-based flora and fauna, especially caribou, but the marine life of Labrador has been the main provider.

Two decades ago, I was at sea in the Okak Islands with an Inuit fisherman who was tending his nets for Arctic char and Atlantic salmon. Several kilometers away, the snow-covered peaks of the Kiglapait Mountains rose out of the sea, crimson in the morning light, while several hundred meters away towered a Greenland type of iceberg, featuring two prominent steeplelike pinnacles. Suddenly, between our boat and the berg, a small whale surfaced. "Grumpus," my Inuit host observed. Such sightings were obviously a common event, hardly worth noting except, in this instance, for my benefit.

By its size and pleated throat, I identified the small whale as a minke. Although the term "grumpus" is sometimes applied to young humpback, sei, or other rorqual species, most often it refers to minke whales, the most common cetacean in northern waters. Fin, humpback, bottlenose, and sperm whales, as well as common porpoise, are also frequently sighted.

Exploitation of the rich resources of the Labrador Sea can be traced to the Maritime Archaic culture, which occupied southern Labrador at least eight thousand years ago, when the ice retreated from most lowland coastal areas. As their name suggests, this ancient Indian people made their living largely from the sea, developing a sophisticated sea-hunting technology, including barbed harpoons and slate spears used to lance sea mammals. Beginning four thousand years ago, a succession of early Inuit cultures, including the classic Dorset culture, moved into northern Labrador from both the Canadian Arctic and Greenland. Shortly after 1300, Thule culture with origins in the western Arctic arrived in northern Labrador from eastern Baffin Island via Resolution and the Button Islands, known locally as the "stepping stones." With their large skin boats and heavy duty harpoon gear, they quickly spread into the rich whale-hunting regions, where they replaced the resident Dorset people who, though largely maritime in their lifestyle, were not fully exploiting the whales and other marine species.

Having retained the maritime skills their Thule ancestors first developed in the western Arctic, contemporary Labrador Inuit have exploited all of the faunal resources from the edge of the landfast ice to the heads of the bays, including marine mammals, birds, and fish. They have also made extensive use of land animals, especially caribou, black bear, and many species of smaller fur-bearing mammals.

*"The Front"*

The presence of landfast or pack ice, open leads, and polynyas has a profound influence on the life cycles of the seals—ringed, bearded, hooded, and harp—commonly found along the Labrador coast, affecting their whelping, molting, feeding, and migrations. Landfast and pack ice is a feature of the coast from November through June. The degree and extent to which landfast ice is formed is influenced by the character of the coast. North of Cape Harrison, there are two types of coastal topography. The first is described as "closed coastal" and is composed of deeply incised bays or fiords, flanked by many islands beyond the headlands. These act to anchor the ice, providing a stable platform for the whelping of ringed seals. The second type is described as "open coastal." It too consists of deeply incised bays and fiords but has relatively few ice-anchoring islands beyond the headlands. This latter type of coastal topography applies to the area north of the Okak Islands, extending to the tip of Labrador at Cape Chidley, and includes the imposing ramparts of the Torngat Mountains—an area with relatively little coastal landfast ice.

Harp seals are the most abundant seal in the North Atlantic and may well be the second most common seal in the world, after the crabeater seal of the Antarctic. As we have seen, there are two breeding populations in the northwest Atlantic: one in the Gulf of St. Lawrence and the Newfoundland stock that breeds at the Front, an area of southward drifting pack ice that forms each spring off southern Labrador and the northeast coast of Newfoundland. The two stocks intermingle and are capable of interbreeding but for the most part remain separate. The stock from the Front is twice as large as the gulf stock, a difference probably related to the available whelping habitat.

Formation of ice at the Front is a complex process. It thickens gradually, transformed from pancake ice and a slushy mixture called "northern slob" into young ice, still too thin to support seals, and finally into the regular winter ice. This first winter ice provides both the necessary stability as well as leads, which the seals require to penetrate to whelping sites. Breathing holes are found in these leads between floes and are kept open by the seals until the ice becomes thick enough for them to climb out onto it. Whelping ice is normally

## GEORGE RIVER HERD

I FIRST saw a George River caribou on the Labrador coast, in late summer, when an Inuit hunter returned from a successful hunting trip near the abandoned northern settlement of Hebron. The next time I saw caribou was on the shores of Hudson Bay, where I was staying with Cree hunters at a spring goose camp. Both of these animals belonged to the George River herd, which at its peak, in the mid-1990s, numbered 800,000 animals. It was not only the largest caribou herd in the world but perhaps the largest herd of free-ranging ungulates on Earth, rivaling African ungulate populations. Yet these animals survive and proliferate in one of the harshest environments on the planet, the boreal forest and tundra of the Labrador-Ungava Peninsula.

The George River herd belongs to the migratory forest-tundra ecotype (a locally adapted variant) of the woodland caribou. They range over a vast area of nearly a half million square kilometers (200,000 square miles), encompassing most of northern Quebec and Labrador, from the Labrador Sea to Hudson Bay. In the past, this herd has undergone wild fluctuations, from very high numbers in the late 19th century to such scarcity twenty years later that biologists contemplated penning some in a zoo to ensure the survival of the species. In 1958, the herd was estimated at only 2 percent of its peak size; today, the herd is again in decline, with the latest estimate pegged at 385,000 animals.

Caribou subsist largely on lichens, which are very slow-growing. Limitations on herd growth may be delayed, however, as the herd compensates by expanding its range, as the George River herd did. Range expansion itself exacts an energetic cost, and eventually food limitations become apparent when new ranges can no longer be colonized. The George River herd has a particularly limited summer range, restricted to the tundra plateaus along the Labrador Sea, 150 kilometers (90 miles) inland from Nain. These are the traditional calving grounds, and caribou concentrate there in part because the dominant winds reduce insect harassment. This summer refuge constitutes only 15 percent of their annual range, and the growing herd has likely damaged this restricted habitat, essentially eating itself out of house and home and partially accounting for the most recent plunge in population.

But caribou numbers around the globe have plummeted by half in recent decades. These drastic declines may be related to global warming, which has boosted temperatures twice as much in the north as elsewhere. As a result, unusual freezing rains in autumn lock away winter forage under ice, the summer plague of flies and insect parasites has increased, and the Arctic "green-up" occurs two weeks earlier, out of phase with the evolved timing of the migration to the calving grounds.

50 centimeters (20 inches) thick and somewhat hummocked. It appears that harp seals prefer this type of ice to thin flat ice, since the hummocks and hollows provide protection for the pups from the wind. The females haul out onto the winter pack ice in late February or early March and give birth several days later. Individual herds usually divide into two main "patches," covering an area of 20 to 200 square kilometers (8 to 80 square miles) and containing as many as two thousand adult females per square kilometer.

The newborn pups weigh about 11 kilograms (24 pounds) and are yellowish. Within three days, their fur turns to a fluffy white, earning them the nickname "whitecoats." Young harp seals are among the fastest growing and most precocious of young mammals. During the brief nursing period of twelve days, they triple their weight on the 45 percent fat content of their mother's milk. Mothers frequently leave their pups during this period but relocate them by odor and perhaps by the distinctive call of the pup. Once abandoned, the pups begin to molt, replacing their white coat with a short silvery one flecked with dark spots. Following weaning, the pups fast for four to five weeks and may lose a third of their body weight. It is believed that this period is a necessary developmental stage, allowing the pups time to develop the physical and behavioral skills necessary for them to forage on their own.

As soon as the females leave the "whelping patch," they are courted by males that have been waiting nearby. The fertilized embryo is not implanted until early August, however, which ensures that the pups will be born at the time of year when ice conditions are likely to be favorable. Ice conditions likely trigger whelping, since in years of reduced ice formation whelping is delayed. Whelping also occurs as the winter ice formation is ending and lasts only long enough to sustain the pups while they grow and become self-sufficient. In years when there is poor ice formation in the Gulf of St. Lawrence, gulf breeders move northward as far as the Front to find suitable ice conditions.

A month after whelping, in early April, both mature and immature seals undergo the annual molt, at the southern edge of the icefields, to replace their fur. Adult males and the young molt first, followed by the mature females. During the molting period, adults rarely eat and may lose up to 20 percent of their body weight. Once the molting process has finished, they depart for their summer feeding grounds, working their way along the Labrador coast and feeding on capelin as they go. Small numbers will summer in northern Labrador, but most will move farther north, arriving along the southwest coast of Greenland in early summer, where they begin to feed heavily on capelin to replenish lost stores of fat. Younger animals appear to move farther northward along the coast as the season progresses, and in fact, most of the summering animals in Greenland are either juveniles or young of the year. Some animals also pass through Hudson Strait into Hudson Bay, penetrating as far as the Belcher Islands, at the top of James Bay. More seals move up the west side of Davis Strait, however, to utilize shore leads such as in Cumberland Sound, on the west side of Baffin Island.

When these inner sounds begin to freeze up in early November, they retreat to southern Cumberland Sound. Numbers build in Frobisher Bay in late November and early December, and by late December both adults and juveniles are again passing through northern Labrador waters, on their way to the whelping grounds.

Harp seals were the basis of an intensive seal industry beginning in the early 18th century. The heyday of the seal hunt was in the early 19th century, when it was not uncommon for a half million pelts to be harvested annually. Most were whitecoat harp seals, though some adult and juveniles were taken as well as a small number of hooded seals. Catches declined in the late 19th century, despite the advent of steam-powered ships. By the mid-20th century, approximately 300,000 harp seals were being harvested annually in the Northwest Atlantic. Increasingly older seals were being taken, resulting in a marked decline in population size and pup production. The population was probably halved to about 1.5 million juvenile and adult seals by 1970. In the 1980s, a European Community ban on seal products, in particular whitecoat pelts, dramatically reduced demand and consequently the annual harvest. As a result, the harp seal population has rebounded and now stands at nearly 7 million older seals, with an annual pup production of a half million animals.

*following spread:* A harp seal pup is known as a "whitecoat." Within twelve days of birth, it will triple its weight, then molt, shedding its namesake coat for a silvery one flecked with dark spots.

Harp seals occur in greater abundance, at any one time, along the Labrador coast than any other type of seal. The ringed seal is the most common resident seal, however, and the only one found along the coast year round. Ringed seals are primarily Arctic and subarctic residents and therefore occur in higher numbers in northern than in central or southern Labrador. They are most numerous along the Labrador coast in late spring, when they occur in areas of landfast ice.

Unlike harp, bearded, and hooded seals, which frequent areas of pack ice, ringed seals breed primarily in areas of landfast ice. Whelping takes place mainly in March and April. The seals give birth in "dens" that they have hollowed out in the sea ice. During the winter they are believed to feed along the floe edge or under the sea ice, traveling between breathing holes, which they

A Greenland-type iceberg towers out of the sea like an Ice Age cathedral.

## BERGS TO BITS

ICEBERGS IN Newfoundland waters are a dramatic feature of the marine environment and a reminder of the island's northerly affiliations. In Labrador waters, there is ice all months of the year. In late August I have seen large icebergs drifting imperiously along the mountainous, island-studded coast as if summer did not exist.

Mountains of lucent, frozen freshwater, icebergs have been compared to floating cathedrals—Ice Age Chartres adrift in the North Atlantic. Icebergs themselves vary a great deal in shape and mass. Large icebergs can be two football fields long, whereas "bergy bits" and "growlers" are a mere 5 to 15 meters (16 to 49 feet) in length. As we know, much of the mass of the iceberg—some 90 percent—is underwater. Large icebergs can weigh more than 10 million tonnes; the smallest bergy bits still tip the scale at over 100 tonnes. The visible silhouettes of icebergs also vary greatly, from round topped "dome" type to the "picturesque Greenland" type, sporting one or more spires.

Most icebergs are calved from the glaciers of west Greenland; others originate in the glaciers of east Greenland and the islands of the Canadian High Arctic, namely Devon, Ellesmere, Bylot, and Baffin Island. It is estimated that the 2-kilometer (1-mile) thick Greenland ice cap produces between 10,000 and 40,000 icebergs every year. In any given year, 500 to 2,500 icebergs drift by the Labrador coast, and of these only several hundred reach the Grand Banks of Newfoundland. Fewer still make it further south. The seabird biologist and writer Richard Brown, in his classic *Voyage of the Iceberg,* observed that from an historical perspective, "there has only been one iceberg." When it ripped a devastating hole in the side of the world's largest ship, the *Titanic,* on April 14, 1912, it was far south—41 degrees north—of where most icebergs ever occur.

A rainbow and an iceberg converge to bring light to a barren coastline.

create when the ice is thin and which they keep open during the winter. When the ice breaks up in spring, they move inshore to feed, largely on Arctic cod and Greenland cod. In summer the population is found farther north along the coast, in bays and fiords.

Bearded seals are not nearly as common in Labrador waters as either ringed or harp seals. Few remain in Labrador waters in summer, and those that do are found near the northern tip at Cape Chidley. They are most frequently sighted in coastal open water and pack ice habitats, and they whelp later than either ringed or harp seals, in late April and early May. Hooded seals also whelp on the pack ice off the coast of Labrador and northeast Newfoundland, but they do so in widely scattered groups, before spending their summers in Greenland waters.

Harbor seals occur in small numbers along the southern and central Labrador coast, where they whelp in caves or on rocky ledges or beaches at the mouths of rivers. Gray seals also occur, though rarely, on the coast, often on outer islands, having wandered far from their breeding grounds in the Gulf of St. Lawrence. Many of the gray seals found in Labrador waters are immature animals.

*Seabirds on the Move*

The Labrador coast is a gathering place not only for marine mammals but for seabirds as well. There are more than a thousand breeding colonies along its largely unsettled length, the largest being the Gannet Islands, which, oddly enough, are named for a ship wrecked there, not the bird, since no gannets nest this far north. It is, however, an important breeding colony for alcids: 17,000 pairs of common murres, 5,000 of razorbills, and 3,700 of puffins. They feed between the islands and the mainland during breeding season, from April to August.

Labrador waters are also a meeting place for seabirds on the move. Common murre chicks, each accompanied by a parent from the great colony at Funk Island, swim to Labrador from late July onward and are joined there by parents and chicks from the smaller Labrador colonies. At the same time, large flocks of thick-billed murres from the huge colonies in Hudson Strait are moving southward to the central Labrador coast, where they will stay until the sea ice forms in January.

The scale of this coming together of the murre clans is truly spectacular. The 7 million adult thick-bills and 1 million commons, together with another million chicks, represent a significant proportion of the breeding murre population of the Canadian Arctic, Atlantic Canada, and west Greenland. Mixed

A fledgling common murre tests its wings before leaving the nest and plunging into the sea.

in with the murres are razorbills—roughly 70 percent of the New World population of this rarest of all the auk breeds on the Gannet Islands. Finally, dovekies, the smallest of the alcids, move into Labrador waters in October from their large breeding colonies in west Greenland and stay until April or May.

The outer edge of the Labrador Current, where temperature boundaries—fronts between the cold current and the slightly warmer Labrador Sea—tend to concentrate plankton, is also an important feeding area for red and red-necked phalaropes. These small marine shorebirds form dense flocks as they pass through Labrador waters on their way to and from their breeding grounds in the Arctic along the entire length of this temperature boundary, from Hudson Strait to the Strait of Belle Isle. Waterfowl also congregate in significant numbers along the Labrador coast, including all three species of scoters: black, surf, and white-winged. Surf and white-winged scoters tend to concentrate in sheltered bays, whereas black scoters often form dense flocks—like thick lines inked on the horizon—farther out to sea. Some fourteen thousand pairs of common eiders also breed along the Labrador coast, and they are joined by Baffin Island eiders during their southward migration.

Finally, in a crescendo to this avian movement, Canada geese and other waterfowl that breed in freshwater habitat in the interior of Labrador—black ducks, common goldeneye, and red-breasted and common mergansers—migrate to the coast to feed at the end of the breeding season.

In his classic book, *Birds of America*, John James Audubon gives us a dramatic bird's-eye view of the Labrador coast from the perspective of the great black-backed gull, which he painted there: "High in the keen air, far above the ragged crags of the desolate shores of Labrador proudly sails the tyrant Gull . . . On widely extended pinions, he moves in large circles, eyeing the objects below . . . onward he sweeps over rocky bay, visits the little islands, and shoots off toward the mossy heaths . . . Far off among the rolling billows, he spies the carcass of some monster of the deep, and, on steady wing, glides off toward it. Alighting on the huge whale, he throws his head upward, opens his bill, and, louder and fiercer than ever, sends his cries through the air."

# 8

# THE ALTERED REALM

## *The Once and Future Atlantic*

IN 1705, A French sealing captain, Sieur de Courtemanche, commented on the morbid, and unaccountable, spectacle of thousands of whale skulls littering the beach near Red Bay, Labrador, on the Strait of Belle Isle. It would be nearly three centuries before the mystery of this marine mortuary was solved, when documents were uncovered by historical geographer Selma Barkham in 16th-century Spanish archives. They told a story of Basques in Terranova—Labrador—where, beginning as early as 1547, this maritime people prosecuted a lucrative whaling industry, exporting the oil to European centers such as Bristol, Southampton, London, and Flanders.

Like most Europeans, the Basques were first drawn to the New World by the bounty of cod, reported by Cabot half a century before. Having pioneered the whaling industry as early as CE 1000 in the northeast Atlantic, the Basques would also have immediately recognized the commercial potential for hunting whales, especially in the narrow strait, where they could be easily pursued from shore stations. The result was North America's first oil boom—and the first large-scale industrial decimation of a wildlife species.

When I visited Red Bay in the early 1980s, the burned blubber of the rendered whales still clung like black lichen to the remains of the stone tryworks, and some of the moldering, moss-covered bones of the great animals were still visible along the shore. Examination of those bones has revealed that they belonged to two species: the Arctic bowhead and

< According to *The Vinland Sagas*, "Wonderstrands" along the Labrador coast is the place where the first Europeans came ashore.

the North Atlantic right whale, both endangered and now seldom if ever seen in the Strait of Belle Isle. Over five decades, it is estimated that the Basque whalers killed many thousands of rights and bowheads, more than enough to precipitate a population decline from which these whale stocks have never recovered.

Before shifting their operations to the other side of the Atlantic, the Basques had already eliminated the populations of right whales in the Bay of Biscay and other nearshore northern European waters, and there are perhaps as few as four hundred right whales fighting for survival today in the Northwest Atlantic. After Red Bay, other New World extirpations followed, as well as the extinction of three species—the great auk, Labrador duck, and sea mink—in the 19th century. More recently, in the last decade of the 20th century, we have witnessed the unthinkable: the commercial extinction of the northern cod stock, the very species that first drew Europeans to North American shores.

### *From the Shore to the Deep Blue Sea*

The new science of marine historical ecology has documented the pattern of exploitation that has led—inevitably, it seems—to the tragic loss of species and wholesale degradation of marine ecosystems on both sides of the Atlantic. This lamentable picture of blind destruction has been reconstructed employing a multitude of sources: the paleontology of soils and marine sediments, the archaeology of prehistoric middens, descriptions of early naturalists exploring the New World, lists of species sold in medieval fish markets, recipes in historical cookbooks, and the logbooks of whalers, to name a few. What has emerged is a very consistent narrative of humans moving inexorably from freshwater to the seashore, then into deeper and deeper waters, leaving behind them a trail of destruction.

This pattern appears early, first during the Roman period, when highly valued oysters were depleted in Italian waters two thousand years ago and subsequently had to be imported from the North Sea. The oyster fishery there finally succumbed to centuries of overexploitation in the early 1900s. On this side of the Atlantic, a similar loss of the once-extensive oyster beds in Chesapeake Bay occurred in the late 19th century.

Heike Lotze of Dalhousie University in Halifax, Nova Scotia, is a world leader in the emerging field of marine historical ecology and has demonstrated that the pattern of land-to-shore-to-deep-sea exploitation has been repeated many times. She has meticulously documented the ecological decline of the world's largest intertidal system, the Wadden Sea, which has one of the most complete records of human exploitation, dating to the Neolithic.

As early as five thousand years ago, the local tribes were making the transition from hunting and gathering to a settled life of farming and fishing. Marine fish gradually became more important as a staple for these coastal people, replacing marine and land mammals, coastal birds, and molluscs. Then, during the Middle Ages, subsistence fisheries evolved into commercial enterprises, driven by market demand from the European inland, where freshwater and anadromous fish stocks like salmon had already been wiped out. By 1800, local haddock, sturgeon, and oyster stocks had been depleted by half, and eventually European sturgeon, Atlantic salmon, shad, sea trout, and the European oyster were driven to extinction in the Wadden Sea. By 2000, the only commercial fishery resources left were shrimp and blue mussels.

This pattern of exploitation and depletion, sometimes ending in extinction, was common to the coastal ecosystems around Europe and by the end of the Middle Ages had driven fishers to the offshore regions in the North Sea and Iceland. Eventually, the loss of local nearshore resources led European fishers to the North American side of the Atlantic, where the patterns—a kind of domino effect—have been repeated on a more compressed timescale but with no less devastating results.

Lotze has applied her probing methodology to this side of the Atlantic with an in-depth analysis of the Quoddy Bay region of southwestern New Brunswick, in the outer Bay of Fundy. Enriched by the outflow of the Saint John and St. Croix River watersheds and driven by marine upwellings, this is one of the most productive coastal regions in eastern North America. As the noted oceanographer A.G. Huntsman once observed: "Nowhere else in the world (this may be questioned) are there waters to compare with those of Passamaquoddy Bay and vicinity, where such vast quantities of fish are taken yearly from a small area." Aboriginal people of the Quoddy Tradition gave it its name, which means "Bay of Pollack." These early residents dined on a cornucopia of marine species, including fishes (cod, pollack, herring), invertebrates (clams, blue mussels, sea urchins), mammals (gray and harbor seals), and a variety of marine birds (common loon, brant, even great auk). While they exploited a diversity of organisms from all trophic levels of the food web over two to three thousand years, their subsistence economy did not result in depletion or extinction of a single species. The consequences of European colonization, however, were in sharp contrast.

The changes to the marine environment came from all directions:

Harvesting of marine resources has systematically exploited the ecosystem, from top predators to primary producers, such as rockweed.

"top-down" through exploitation, "bottom-up" from nutrient loading leading to oxygen depletion (or eutrophication), and "side-in" by habitat destruction and pollution. Exploitation began with whaling and a seasonal cod fishery in the 16th and 17th centuries but upon settlement of the region in the 18th century quickly expanded to include most marine species, of which herring and lobster were among the most important. The decline of groundfish species in the 1960s and 1970s—and their eventual commercial collapse in the 1990s—led to a practice that has been called "fishing down," whereby fishing is first directed at long-lived, fish-eating species such as cod and then, when they have been depleted, to ever-lower trophic levels: first to plankton-eating fishes such as herring and then, as in the Quoddy region, to invertebrates such as shrimp and crab, and finally, in the 1980s and 1990s, to rockweeds, the primary producers. In effect, this amounts to a sequential mining of marine resources—going ever lower in the food chain and stripping out each level of the system—which ultimately can lead to major changes in the structure of the marine food web itself.

The changes wrought have been both quantitative and qualitative. Today's groundfish catches are a fraction—3 to 37 percent—of what they were one hundred years ago. Not only did dragging the seabed with otter trawls scoop up more fish than traditional hook-and-line methods, but the heavy gear also destroyed seabed habitat. Scientific surveys now show that the numbers of desirable groundfish, such as cod, pollack, and haddock, have remained low for the last thirty years, while the numbers of noncommercial species, such as spiny dogfish, longhorn sculpin, and thorny skate, have steadily increased. After the collapse of the groundfish, their prey species—shrimps, crabs, lobster, and sea urchins—also burgeoned, signaling a shift in the structure of the marine food web.

Anadromous fishes fared even more poorly because of the damming of their rivers of birth, such as the St. Croix, and the indirect, "side-in" effect of pollution. Salmon had nearly been extirpated from the system by 1909. Although effective fishways enabled salmon to return to the river in the 1980s, these populations, along with all of those from rivers in the inner Bay of Fundy, have since collapsed, perhaps because of high mortality in the sea from unknown causes. Today, the return of alewives to the river, estimated at a potential 31.7 million fish, is less than 1 percent of that total. Nutrient enrichment, leading to oxygen depletion, has occurred here, as elsewhere along the eastern seaboard, notably in Chesapeake Bay, with a decline of long-lived rockweeds and the more frequent appearance of toxic phytoplankton—"red tides"—killing fish and contaminating shellfish beds.

Despite this long litany of depredations, the Quoddy region remains relatively productive, though much less so than it was historically. Some severely depleted species have been able to make a slow comeback with the implementation of protection for seabird colonies, the cessation of whaling, and the removal of bounties on seals, for example. But other severely depleted species' recovery may have to be measured in decades and perhaps centuries, as seems to be the case for the critically endangered North Atlantic right whale, which has found a last refuge in the North Atlantic in the outer Bay of Fundy.

### *Top to Bottom: Cascading Ecological Effects*

The picture globally, especially for large marine species, is far from encouraging. A survey carried out by Boris Worm and the late Ransom Myers—like Lotze, both members of Dalhousie's Biology Department—revealed that large predatory fish communities had been depleted by at least 90 percent in the last fifty to one hundred years. This disturbing revelation applied to marine

## DEEP-SEA CORALS

A SURPRISING discovery of the last two decades has been extensive deep-sea coral beds off the North Atlantic coast. Normally, coral is associated with warm, shallow tropical waters. Two-thirds of all known coral species, however, live in deep, cold, dark waters. Almost thirty species of corals have been identified in Atlantic Canada alone.

The highest diversity of corals is found along the edge of the continental shelf between the Gully and the Laurentian Channel at the edge of the Scotian Shelf. One of these, discovered only in 1992, 150 kilometers (90 miles) offshore of Canso, Nova Scotia, is the reef-forming species *Lophelia pertusa*. Some *Lophelia* reefs may be many thousands of years old, and the largest cover several square kilometers and rise 20 to 30 meters (65 to 100 feet) above the sea floor. Other deep-sea corals are either solitary or form tree-like structures with a main stalk and branches. One of the latter is a gorgonian coral known as bubblegum coral for its gaudy pinkish color. Growing less than 2.5 centimeters (1 inch) per year, these large bubblegum trees may be five hundred years old. Ten species of these gorgonian corals are found in the deep waters of the Scotian Shelf, of which nine are found in the area of the Gully. A large aggregation of gorgonian corals also lies in a deep channel between Browns Bank and Georges Bank, 150 kilometers (90 miles) south of Nova Scotia.

Although they look like plants, corals are actually colonies of small animals related to sea anemones. Each coral polyp is an individual animal that secretes a skeleton upon which the colony forms and lives. The gorgonian corals, or "tree corals," are also known as sea fans or "horny corals," in reference to the pliable skeleton that the polyps secrete. Most corals need hard substrates to anchor themselves and, as filter feeders, strong currents to deliver their food, which consists of plankton and detritus. Both of these conditions are met along the edge of the Scotian Shelf.

The corals themselves provide an attractive habitat for other marine species, including fish, molluscs, crustaceans, and sponges. Different fish species appear to be associated with specific species of coral. Off Nova Scotia, fishers know that haddock are found around *Gersemia rubiformis,* a soft coral known as "strawberries," while cod and halibut frequent bubblegum tree coral. Unfortunately, dragging or bottom trawling has destroyed many corals, pulverizing communities that may be hundreds of years old. It will take that long again for them to recover, if at all.

predators such as sharks, tuna, billfish, and large groundfish like cod, as well as marine turtles and cetaceans. Until 2003, when Myers and Worm published their landmark paper in *Nature*—"Rapid worldwide depletion of predatory fish communities"—the 19th-century notion that marine life was inexhaustible had pretty much held sway, at least in the public mind. It had been assumed that marine habitats were too remote and fish populations too fecund ever to be exhausted, but the statistics told a different story. The United Nations Food and Agriculture Organization concluded that two-thirds of the global fisheries were fully exploited, overexploited, or depleted by the turn of the millennium—a number that has since increased to 80 percent.

Recent studies have shown that large predatory fish like bluefin tuna have been depleted by 90 percent in the last fifty to one hundred years.

For the entire North Atlantic, large fishes have declined to a mere 11 percent of their numbers in 1900. The fate of large sharks in the Northwest Atlantic was even more disturbing, as they have been reduced to a mere 2 percent of their former abundance in the past 50 years.

Declines have occurred with astonishing speed; community biomass was reduced by 80 percent within fifteen years after industrialized fishing began in any given sector of the sea. At this rate it is possible for a large fish to go extinct, or virtually so, before anyone can make a move to protect it or even notice that it is in peril. A case in point is the barndoor skate.

As the name might imply, the barndoor skate was the largest skate in the Northwest Atlantic and once was found on the continental shelf from the Grand Banks of Newfoundland to North Carolina. Henry Bigelow describes it in his encyclopedic *The Fishes of the Gulf of Maine* as "growing to a length of 5 feet," and possibly reaching 6 feet, "though there is no definite record of one that large." He says that it was "plentiful off the outer Nova Scotia shore" and abundant on Georges Bank, where an experimental trawl would routinely dredge up twenty of them. It has been estimated that there were 600,000 individuals in the 1950s, a number that shrank to 200,000 in the 1960s, and plummeted to a mere 500 individuals in the 1970s. Recent surveys, however, indicate that barndoor skates have survived in small numbers only in the very deep waters off Newfoundland, as well as on Georges Bank and Browns Bank, probably because of the faster growth rate there (due to warmer water temperatures) and the seasonal closure of these latter banks to trawling. The authors warn, however, that if a fish as conspicuous as the barndoor skate can nearly disappear, virtually unnoticed, in an area regularly surveyed, then the fate of less well known species might well be worse.

BARNDOOR SKATE

The demise of these top predators is cause for alarm in itself, but more so because each loss has top-down cascading effects on the marine food web. The classic example comes from the Pacific, where overexploitation of sea otters in Alaska led to population explosions of their sea urchin prey, which in turn caused an overgrazing of kelp forests. Likewise, the recent collapse of groundfish stocks in the Northwest Atlantic due to overfishing cascaded through the food web, demonstrating once again that all parts of the marine ecosystem are connected, even on the high seas. On the Scotian Shelf, plankton-eating fishes such as herring, capelin, and sand lance—once the prey of cod and other bottom-dwelling fishes—burgeoned in numbers when released from predation pressure, and the Newfoundland cod collapse led to a similar dramatic increase of shrimp and crab and possibly lobsters as well. In fact, Worm and Myers showed that on both sides of the North Atlantic, where cod numbers were depleted shrimp numbers rose, suggesting that there was top-down control of prey numbers in these ocean systems.

One of the best examples of the cascading effects of top predator removal comes from the Mid-Atlantic. Surveys carried out off North Carolina since 1972 reveal drastic declines in the number of great sharks, which are targeted for their fins (a luxury item in Chinese cuisine) or reeled in as by-catch in the tuna and swordfish fishery. Declines in seven great shark species range from 87 percent for sandbar sharks to 99 percent for bull sharks and, in Chesapeake Bay, a 99 percent decline in tiger sharks between 1974 and 2004. These great sharks prey on smaller sharks, rays, and skates. With these great predators out of the way, their prey proliferated. In particular, cownose ray numbers jumped by an order of magnitude—to more than 40 million. As they migrated south from northern estuaries each fall, they stopped over in North Carolina sounds, where they sought out their preferred prey, the bay scallop. By 2004, cownose rays had almost wiped out the bay scallop population, bringing to a close North Carolina's century-old bay scallop industry.

It was another example of the interconnectedness of marine ecosystems and a dramatic demonstration of how the removal of one species echoes through the system with unexpected consequences—economic and ecological—down the line. The top-down effects can ultimately impair the recovery of the predator population through a predator-prey role reversal. This appears to have happened in the southern Gulf of St. Lawrence, where the recovery of the collapsed cod population may be held in check by the predation of cod eggs and larvae by the now more abundant prey species, herring and mackerel.

The human response to this disruption in the marine ecosystem has often been to continue fishing down through the food web: first taking out the top predators, then concentrating on their prey, and finally harvesting the plants such as rockweed that are at the base of the food-production pyramid. In the last two decades a euphemism has emerged in bureaucratic circles to describe and justify this practice: the term is "underutilized species," which Ransom Myers, with typically trenchant humor, once ruefully characterized as government-speak for "some left."

Current policy is an endgame with potentially disastrous consequences for our ability to feed ourselves and for the recovery of the marine environment to some semblance of health. In a now famous paper published in the prestigious journal *Science* in November 2006, an international group of ecologists and economists, with principal author Boris Worm, warned that if the stripping of the oceans continued at the same rate as over the last fifty years, all species of wild seafood currently being exploited could collapse by the year 2048. Co-author Steve Palumbi, of Stanford University, observed that unless we begin to "manage all the oceans' species together, as working ecosystems, then this century is the last century of wild seafood."

It is a sobering wake-up call. Society will lose not only a vital source of food and pharmaceuticals but also the other services the ocean provides, such as flood control and waste detoxification by oyster reefs, sea grass beds, and coastal wetlands. The paper also poses potential solutions: the key to preserving these vital services is to maintain biodiversity. Restoring biodiversity has been shown to increase productivity fourfold, and fish diversity also aids the rate of recovery of collapsed populations. Evidence for the link between biodiversity and the productivity and resilience of marine ecosystems comes from examining fully protected marine reserves and large-scale fisheries closures. In these cases, species diversity of both target and nontarget species has been boosted by nearly a quarter (23 percent), and these increases in biodiversity have translated into large increases in fisheries productivity around reserves. The rebound time can be surprisingly quick—three to ten years.

*Historical Amnesia and Future Shock*

University of British Columbia fisheries scientist Daniel Pauly has hypothesized that marine scientists generally suffer from a historical amnesia that has resulted in a misguided perception of what a productive marine ecosystem actually looks like. Most often, he says, it is based on already degraded systems, and so our assumptions are seriously shortsighted. Historical records now show that many highly valued marine animals—great whales, sea turtles, seabirds, and large fishes like tuna—were severely depleted before the midcentury and that many reached their low point decades to more than a century ago. Historical trends of some 256 records reviewed by Lotze and Worm followed similar trajectories: slow changes over millennia, followed by rapid depletion over the last 150 to 300 years, with some recovery in the 20th century, particularly for marine mammals and birds.

Although exploitation and habitat loss have been the main culprits in the historical declines in population observed in marine species, climate variation has played a large role in long-term fluctuations. Cod catches in Newfoundland from 1505 to 2004 indicate that climate-related declines occurred from 1800 to 1900 during the Little Ice Age, large declines in the 1960s due to overfishing, and the collapse in the late 1980s perhaps caused by both. In this age of global warming, climate change may become more critical to predicting the future of marine ecosystems.

Overfishing was the main cause of the dramatic ecosystem shift observed in the Northwest Atlantic in the early 1990s, from one dominated by fish-eating species such as cod to plankton-consuming ones like herring. Climate change, however, also seems to have played a significant role. In the late 1980s and early 1990s, climate change over the Arctic Ocean began to boost the outflow of low-salinity water from the Canadian Arctic Archipelago into the Labrador Sea. The freshwater had its source in the increased melting of permafrost, snow, and ice in the Arctic, as well as increased precipitation in the last three decades. This pulse of low salinity water passed Georges Bank and reached the Mid-Atlantic Bight by 1991, causing a freshening of the seawater along the entire Northwest Atlantic Continental Shelf.

As a result, a larger and deeper than usual layer of freshwater now sat atop the seawater, which is less buoyant than freshwater. These layers are normally mixed to a depth of 25 to 200 meters (80 to 650 feet) in summer by wind turbulence. But in autumn, as the air temperature cools, the temperature and density differences between the surface layer and the cooler, saltier layer below break down, increasing the mixing of the layers. With the influx of freshwater due to Arctic climate change, however, the surface layer is remaining buoyant, with major biological consequences. Usually, as the mixed layer deepens during autumn, phytoplankton numbers decrease as they spend less time near the surface, where they are exposed to sunlight necessary for their growth. Now phytoplankton in the surface layer remain abundant, fueling the growth of zooplankton populations, which feed on the tiny plants. The inflated zooplankton numbers are probably another reason that herring became much more abundant in the 1990s.

The freshening of the waters in the Gulf of Maine has also altered phytoplankton and zooplankton assemblages in the gulf, with a general change from large zooplankton to smaller zooplankton. These changes may require animals higher in the food chain to forage for longer periods of time to meet their energetic needs. *Calanus finmarchicus* determines the arrival time of migrating North Atlantic right whales in the spring and their reproductive success. Changes in the magnitude and the timing of the peak abundance of the tiny prey species may alter migration, behavior, and, ultimately, numbers of this critically endangered whale.

According to the 2010 State of the Gulf of Maine Report, the higher silicate concentration found in the fresher Labrador Current waters is expected to favor diatom production, which is thought to increase overall ecosystem productivity, since diatoms are larger than dinoflagellates.

Atmospheric warming in concert with melting sea ice is changing the physical oceanography as well as the ocean chemistry of the Northeast U.S. Continental Shelf, including the Mid-Atlantic Bight and the Gulf of Maine. Climate largely determines an ecosystem's structure, function, and productivity, and surface sea temperatures have been rising in the region for the last forty years and are predicted to rise another 2° to 4°C (36° to 39°F) by 2080.

Temperature is critical to the growth, development, and survival for cold-blooded species found in coastal and marine waters. It is also key to where organisms are found and when and where they make their migrations. Along the Northeast shelf, the thermal habitat between 5° and 15°C (40° to 60°F) has decreased in the last twenty years, while at the same time the coldest and warmest habitats have been increasing. These trends have produced a habitat "squeeze" for most of the species in the North Atlantic.

Warming seas are driving them to colder waters. Over the last forty years about half of the fish stocks studied in the Northwest Atlantic have shifted their center of biomass (where they are most abundant) northward or have moved to deeper depths. As an example of the latter, the distribution of

A traditional swordfish boat pursues its prey at the surface, while deep-sea species such as ocean perch, or redfish, are trawled from the sea bottom.

Atlantic surf clams has shifted to deeper, cooler water in response to increased coastal temperatures. Red hake has moved its central range northward as waters have heated up, and so its numbers have decreased in the Mid-Atlantic Bight and are now concentrated in the western Gulf of Maine. The restricted shape of the Gulf of Maine limits the north-south movement of species, and as a result, half of the thirty-six fish stocks examined have also moved to deeper waters to avoid the warming water temperatures.

Warming sea surface temperatures are thought to be the cause of a decrease in groundfish and an accompanying increase in pelagic fishes and benthic invertebrates in Narragansett Bay in southern New England. Similarly, they have had a dramatic negative effect on sea grass habitat in coastal New Hampshire. Loss of sea grasses has also been attributed to changes in the timing of bird migrations and storm patterns related to global warming.

The Gulf Stream is pushing farther north into the Gulf of Maine. Overall, the changes in the fish and invertebrate communities in the Gulf of Maine make it look more like southern ecoregions did in the past. All along the Northeast U.S. coast, warm-water species are now more dominant than cold-water species. Although warmer waters boost the growth of adult cod, they will negatively affect survival of cod in early life stages, and therefore the cod population is expected to decline.

Climate change can also affect the life cycles and migration patterns of local species far from the coast. American eel populations are in decline because of the mortality of eel larvae attributed to changing sea temperatures and wind conditions in the Sargasso Sea. Similarly, the failure of juvenile salmon (smolt) to survive once they leave Maine rivers and begin their ocean migration is thought to be related to climate change factors.

Human communities will also be greatly influenced by climate change. Exploited resources, such as fish, are the historical basis of many coastal communities, and protected species, such as whales, currently attract large numbers of people to coastal regions for tourism and recreation. Native peoples are also dependent on these resources for preserving their culture and traditions. Threats to these resources have economic and cultural consequences for coastal communities.

The U.S. Global Change Research Program has predicted a rise in precipitation of as much as 25 percent by 2100, with increased flooding from storms, rising sea levels, and coastal land loss. The transportation, communication, energy, water, and waste disposal systems of the major Northeast cities will all be put at risk. Sea level rise will also exacerbate stresses to estuaries, bays, and wetlands from increasing pollutants, temperature, and salinity. Additionally, sea level rise will allow storm surges to reach farther inland, and climate change could cause an increase in the intensity and frequency of storms in the Northern Hemisphere. Low-lying islands could be submerged, and areas where human infrastructure prevents landward migration of coastal ecosystems will be at particular risk.

Plans to protect ecosystems and habitats will include upgrading sewage treatment and stormwater runoff systems and protecting wetland habitat. In order to mitigate the impact of climate change on marine ecosystems and the people and communities who rely on them, monitoring programs to assess changes in the status of migratory fish, invasive species, harmful algal blooms, and other coastal resources will have to be developed and maintained.

The Northwest Atlantic ecosystem is being altered both by top-down forces, such as overfishing, and bottom-up factors like climate change. Some of the changes observed would likely have happened even without the added stress of overfishing and are likely to increase in the 21st century, even if greenhouse gas emissions are controlled. All marine life will be dramatically affected by the changes already set in motion. Seabirds and cetaceans are two prominent groups, besides fish populations, that are likely to be affected and, in some cases, already have been. The warming atmosphere will cause an accelerated rate of rise in sea level by melting ice caps, which will pour more meltwater into the oceans, and through thermal expansion of the water

## THE TRANSFORMATION OF BOSTON HARBOR

HUMANS HAVE been called a littoral species, clinging to the coast as tenaciously as other shoreline organisms. Unlike the presence of other organisms, however, our presence has often resulted in the degradation of the inshore environment. Pollution has been especially intensive near highly populated urban areas, which have been the source of both industrial and domestic wastes dumped into the oceans, on the doorsteps of our cities.

A classic case is Boston Harbor, which, twenty years ago, was known as the dirtiest harbor in America. Fish and crabs from the harbor exhibited signs of disease, such as fin rot and ulcers. In 1988, the Massachusetts Department of Public Health issued a health advisory warning citizens not to eat fish and shellfish from Boston Harbor, including Quincy Bay. They were also ordered to stop eating lobster tomalley—the soft green liver and pancreatic tissue in the body cavity of the cooked lobster, considered a delicacy by some—because of abnormally high levels of chemical contaminants, including polychlorinated biphenyls (PCBs), which have been linked to cancer in humans. Such contaminants accumulate in the fat tissues, and their concentrations increase progressively through the food chain, a process called biomagnification. As a result, predators and scavengers at the top of the food chain (fishes, marine mammals, and seabirds, in the oceans) accumulate the highest and potentially most toxic levels of contaminants.

Pollution problems arose in Boston Harbor long before industry began pumping synthetic organic chemicals and heavy metals into the environment. Sewers funneled household wastes and stormwater runoff into the harbor as early as the 17th century, with predictably disastrous results. A severe cholera outbreak in the 1860s led to an upgrade of the sewer system, but the burgeoning coastal population and the growth of industry in the first half of the 20th century overwhelmed the antiquated treatment plants, further degrading the health of Boston Harbor. Similar scenarios played out in urbanized coastal areas such as New York.

Despite further upgrades, the environmental quality of the harbor continued to decline and reached a breaking point in 1985 when the city was successfully sued for violation of the Clean Water Act of 1972. The crisis led to the creation of the Massachusetts Water Resources Authority and a $4.5 billion overhaul of the system and clean-up of the harbor. Treated sewage is now piped 15 kilometers (9.5 miles) from the Deer Island plant into Massachusetts Bay.

All treatment plant discharges to Boston Harbor ended in September 2000, and the positive results have been dramatic in the last decade. Thirteen kilometers (8 miles) of beaches once closed to recreation because of bacterial contamination have opened to swimmers, and native fish species, such as smelt, herring, striped bass, and bluefish, have returned to the harbor. Harbor porpoises and seals are again seen around the islands, and the lobster and shellfish industry now contributes more than $10 million to the local economy. There have been significant decreases in the levels of contaminants, including heavy metals such as lead, and chlorinated compounds in the sediments in Boston Harbor have significantly decreased, as have the incidence of tumors and the bioaccumulation of contaminants in fish and shellfish. At the same time, minimal contamination has been observed in Massachusetts Bay.

In the last two decades, Boston Harbor has been transformed from the dirtiest in America to "a great American jewel."

Today, the harbor has been transformed from the dirtiest harbor in America to what the Environmental Protection Agency has called "a great American jewel."

Despite four centuries of relentless exploitation, the Atlantic still has the capacity to renew itself and inspire awe.

column itself. Together, the changes are projected to cause a global increase in sea level of 1 meter (3 feet) by 2050.

This scenario will have two major impacts on seabirds. There is likely to be a northerly shift in their ranges, probably in response to shifts in the ranges of prey species. Such a shift has already been observed with northern gannets. During the last century, gannets have expanded their range northward. Colonies at Cape St. Mary's and Funk Island, for example, have grown as water temperatures have warmed up and the gannets' prime prey, mackerel, has moved north. Rising sea level, in combination with greater storm surges and tides, may wipe out important low-lying breeding colonies like those at Funk and Sable Islands, as well as shrinking the great mudflats in the inner Bay of Fundy, which are critical feeding grounds for migrating shorebirds. Changes in tidal range may also disrupt the spawning cycle of horseshoe crabs in Delaware Bay, putting it out of phase with the red knots' migration patterns in spring.

Conditions might well improve for a number of seabird species. Low Arctic breeders like common murres are likely to expand their breeding colonies northward, as will southern non-breeders like shearwaters, which use the northwest Atlantic as a feeding ground during the austral winter. Winter visitors from the Arctic will also retreat northward as the Atlantic warms. Ivory gulls now scavenge the herds of harp seals and hood seals as they pup at the Front off southeast Labrador. With warming seas, seals will move north and the gulls with them. Breeding conditions in the north for low and high Arctic breeders like thick-billed murres and dovekies might well improve with warming seas, opening up new feeding areas in spring with the reduced ice cover and exposing more potential breeding sites as melting snow cover exposes more seaside slopes.

Rising sea level will negatively affect several types of marine bird habitat. Coastal sandbars on Nova Scotia's Atlantic coastline and the Gulf of St. Lawrence will be lost and with them the breeding habitat for endangered piping plovers and many terns. If Sable Island itself disappears under a warming and rising sea, the rare Ipswich sparrow will almost certainly become extinct, and the world's largest colony of gray seals will be lost. Increased tides at the head of the Bay of Fundy will also put pressure on the extensive saltwater and freshwater marshes, which are important breeding areas for "puddle ducks" like the ubiquitous black duck as well as rarer species in Atlantic Canada such as northern shoveler, gadwall, and American widgeon.

In pondering the consequences of warming and rising seas, the late seabird biologist and author Richard Brown wrote that "it is hard to know how much to believe of these scenarios." With insulating feathers and subcutaneous fat, seabirds are well adapted to cope with extreme cold, but some have greater difficulty in keeping cool—for instance, northern gannets, which find it harder to deal with heat stress than with the cold. So, Brown concluded, "warm temperatures may therefore set southern limits to seabird distributions that need have nothing to do with availability of suitable prey or breeding sites."

The projected picture with cetaceans is different again. Deep-water whales are mobile, have large ranges, and face few barriers to their movements. They would most likely adapt readily to system-wide changes by moving to parts of the ocean that are more favorable and therefore would be animals of least concern related to climate change. Long-term studies of deep-water whales, like sperm and northern bottlenose, and dolphins in the Atlantic, Pacific, and Indian oceans, predict that in response to the global change scenarios put forward by the Intergovernmental Panel on Climate Change, cetacean diversity would decline across the tropics and increase in higher latitudes. But in the Sable Island Gully, productivity peaks at temperatures of 12° to 16°C (53° to 61°F), then decreases substantially at higher temperatures. In addition, populations of cetaceans that have limited ranges or specialized habitat requirements—like Arctic species such as narwhals, bowhead, and belugas, or baleen whales that migrate to polar regions to feed—may be compromised by the rapidly changing conditions in the north.

Overall, global warming seems to predict an increase in diversity in temperate regions of the oceans and a decrease in the tropics, as is the case on land. But as Richard Brown pointed out, "we must not underestimate the unpredictability of the animals themselves, however well we think we know them."

He gave the example of the capelin, which today can be seen spawning on the gravel beaches around Newfoundland. During the last glaciation, eighteen thousand years ago, there was even more spawning habitat as the southern tip of the Grand Banks—that "vast apron"—emerged. His Neolithic counterpart, Brown surmised, might well have predicted that the capelin stock and the predators dependent on it would be decimated as these beaches disappeared under the waves and the ocean climate warmed as the Gulf Stream moved north. "Yet," Brown pointed out, "the capelin still spawn there, 375 kilometers (230 miles) offshore, on 'beaches' 50 meters (160 feet) underwater." And this offshore stock of capelin—separate from the inshore stock—supports molting greater shearwaters, lactating baleen whales, and, until very recently, a major groundfish population.

*Back from the Brink*

Despite these grim scenarios, some modern-day examples give hope that timely corrective human action in concert with the resiliency of natural systems can pull us back from the edge of disaster.

According to George A. Rose, a warming ocean should benefit Newfoundland and Labrador cod stocks, which live near the northern limits of the range of the species. In these ecosystems, cod are limited more by cold water than by warm. If warming does occur, cod will move farther north on the Labrador coast, following capelin, which will move north first, as has already happened in Icelandic and Greenland waters.

There are other hopeful signs. A paper published by Worm and his colleagues in *Science* in 2009 showed that, since the 1990s, fishing pressure has eased up in a number of well-managed regions globally, including Newfoundland and Labrador and the Northeast U.S. Shelf. Although these measures, including the cod moratorium in Newfoundland, have only been instituted in the wake of overfishing and stock collapse, there are encouraging results.

The rarest of the world's great whales, the North Atlantic right whale, seems poised to make a comeback due to recent conservation measures.

Haddock stocks have substantially rebuilt on Georges Bank and the Northeast U.S. Shelf, and there is now evidence of cod recovery in Newfoundland, where stocks have increased by 70 percent on the Grand Banks since 2007. Worldwide, the key to success is to combine protected areas with other management tools, such as fishing gear restrictions, lower catch quotas, and incentives for fishers to support the conservation of stocks. Taken together, these measures provide promise that despite a long history of overexploitation, marine ecosystems can still recover.

The North Atlantic right whale is another case in point. Nearly hunted to extinction, first by the Basques and later by the Nantucket whalers, this most endangered of the great whales is finally showing signs of recovery. Protected from hunting since the 1930s, the population failed to rebound as expected, with as few as three hundred animals in existence when they were discovered in the 1970s to be concentrated on feeding grounds in the outer Bay of Fundy every summer. It was feared that so few females had survived that the

## COASTAL ACROBATS

ATLANTIC COASTAL waters, from Long Island southward, are home to bottlenose dolphins, relatively small, toothed cetaceans which frequently can be seen from shore performing their aquatic acrobatics. The coastal bottlenose populations are separate from offshore stocks, which are generally larger and prefer deeper waters.

The coastal population consists of three principal groups. A northern migratory stock moves northward in the summer, as far as Long Island, then retreats to the Cape Hatteras region for the winter. The southern migratory stock summers in nearshore North Carolina waters, then moves along the south Atlantic coast in winter. Five coastal resident stocks consist of animals that do not migrate but instead occupy estuarine habitats along the coast, such as Pamlico Sound, year-round. However, some of these estuarine stocks move onto the coast during the winter and overlap with the northern migratory animals.

The two coastal stocks seem to be separated on the basis of water temperature. They travel in small groups, or pods, of approximately ten animals. Total population numbers are not known; estimates vary from twelve thousand to as few as two thousand. Annual mortality due to encounters with fishing gear is at least 10 percent in the northern and southern migratory stocks. From 2002 to 2006, 1,570 bottlenose dolphins were stranded along the Atlantic coast from New York to Florida. The cause was not always apparent, but 32 percent of the dolphins showed evidence of entanglement in fisheries gear.

In the summer of 1987, an unprecedented number of dead and dying dolphins began washing up on the coast of New Jersey, and by the spring of 1988, a total of 750 were accounted for along the coast south to Florida. It is estimated that more than half of the population of Atlantic bottlenose dolphins may have died during this period. Studies of the die-off, organized by the National Oceanic and Atmospheric Administration (NOAA), indicated that the dolphins had died by eating fish tainted with a naturally occurring toxin caused by "red-tide algae."

Red tides occur when environmental factors—including the right water temperatures, nutrient supply, and sufficient sunlight—combine to allow an explosive growth of dinoflagellates, single-celled plankton that, in sufficient quantities, literally turn the waters red.

One of the planktonic organisms in red tides is a species called *Ptychodiscus brevis,* which produces a neurotoxin known as brevetoxin. This toxin is believed to have killed the dolphins when the red tide organisms were swept into the Atlantic from the Gulf of Mexico and ingested by planktivorous fishes such as menhaden. When the dolphins consumed the fish, they were fatally poisoned, or

became sick and succumbed to bacterial infections. Autopsies of the carcasses also yielded very high concentrations of persistent organic pollutants such as polychlorinated biphenyls (PCBs). The levels ranged from 13 to 620 parts per million (PPM), with one animal registering 6,800. Currently products are required to be labeled as toxic waste if they contain 5 PPM. It is not known whether these levels of contaminants contributed to the deaths of the dolphins, but animal studies have indicated that PCBs can compromise immune systems.

Bottlenose dolphins belong to three principal populations that occupy coastal waters between Long Island and Florida.

population had gone through a "genetic bottleneck" and was therefore limited in its reproductive capacity.

But it became apparent that a second factor limited their natural increase—high mortality due to ship strikes and entanglement in fishing gear. As a result, international shipping lanes in the Bay of Fundy were moved in 2003, and fishers both in Canada and the United States are being taught how to prevent whales from getting caught up in their nets, traps, and lines. Even so, as recently as 2001, only one right whale calf was produced on the birthing grounds along the southeast coast of Florida. But the very next year, thirty-one calves were born, and in 2009 a record thirty-nine calves entered the population. It appears that an abundance of zooplankton in the Bay of Fundy is contributing to the health of adult whales, which are then more likely to procreate and reproduce.

This population, which was the target of the first industrial "fishery" in the New World, may finally be on the sea road to recovery, providing hope that other severely depleted species might also be saved.

### *Home Waters*

Last summer I boarded the ferry in Blacks Harbour, New Brunswick, for the two-hour passage to Grand Manan Island. It was a brilliant summer day, a gentle sea breeze moving the waters and a silvery light glancing off the wave crests. Astern, a large, heart-shaped herring weir was silhouetted in the harbor, and we were hardly away from the dock when I saw the first spouts of whales chasing herring along the shore. "A family of fin whales," the ferry deck hand said. "Been here all summer."

We were entering the Grand Manan Basin at the mouth of the Bay of Fundy. This section of the Northwest Atlantic, where the famous Fundy tides drive the food production system, is the last refuge of the North Atlantic right whale, but it is also home and meeting place for a number of other cetacean species, including the fin whale, the second-largest of the great whales, and, at the other end of the scale, the relatively diminutive harbor porpoise. I soon spotted the black backs of these porpoises as they too pursued shoals of herring.

Other fishers were at work. On the horizon were two fishing boats, long-lining for cod and pollack. Gannets were soaring on their great black-tipped wings and diving headlong into the waves, sending up geysers of their own. Cruising between the wave troughs were scores of shearwaters—greater and sooty—seabirds that annually flee winter in the South Atlantic to take advantage of the abundant food resources in the bay. Small, dark flocks of auks—razorbills or puffins, hard to tell apart in the strong light—and sea ducks also rode the waves on feeding forays from their nearby breeding islands.

These were also my home waters, I realized, for I had grown up just across the bay on the Nova Scotian shore. This gathering of marine life—herring, whales, and seabirds, underpinned by an invisible blossoming of microscopic phytoplankton and zooplankton—had nurtured my early interest in the natural world and my long-term commitment to its preservation. Seeing this marine bounty on display once again stirred the same ecstatic feelings that I had experienced at St. Vincent's Beach in Newfoundland, when I watched through the natural aquarium window of a breaking wave as humpback whales rushed ashore to gulp down capelin trying to beach themselves in an orgy of renewal.

A week later, I found myself in the inner Bay of Fundy, on the shores of Minas Basin near Wolfville, where I had taken my degree in biology at Acadia University. I drove past the Grand Pré National Historic Site, which commemorates the tragic Expulsion of the Acadians in 1755. One branch of this New World diaspora became the Cajuns of Louisiana. Here, their legacy lives on in the fertile dykelands that they created from the salt marshes three and a half centuries ago by holding back the sea with an ingenious system of earthen

A herring weir stands as a symbol of a productive past and a hopeful future for the Atlantic.

A sooty shearwater dances across the waters of the outer Bay of Fundy, where it comes to escape the austral winter.

dykes and drainage devices called *aboiteaux*. The still-productive fields of ripe corn and grain were beginning to turn an autumnal gold.

I was on my way to Evangeline Beach, named for the heroine of Longfellow's poem about the Expulsion, where I was hoping to see shorebirds. Here, they stop over to fuel up for their migration to South America, on the invertebrate riches of worms and mud shrimp secreted in the tidal flats. It is one of two sites in the inner bay that have been declared a Western Hemisphere Shorebird Reserve.

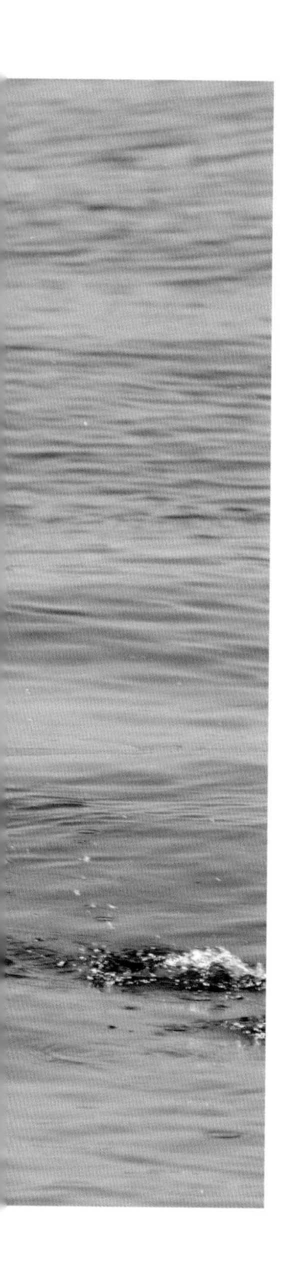

The tide was high, lazily lapping the shore. Scanning the pearl-gray waters of the Minas Basin, I could see a dark line midway between myself and the distant headland of Cape Blomidon. At first I thought it must be a powerful tide rip, but there was something about the way the inky scrawl moved that seemed alive. I raised my binoculars, and the source of this commotion came into focus—tens of thousands of sandpipers, in massive bands, crisscrossing and reversing direction like the shuttles of a living loom. Birds were strung out along thirty degrees of the horizon, perhaps the largest flock that I had ever witnessed, surely numbering more than a hundred thousand.

*following spread:* The epic migrations of shorebirds knit together the hemispheres, north and south.

In a week these birds, having doubled their weight, would fly 4,000 kilometers (2,500 miles) nonstop over the Atlantic to the north coast of South America, employing the same trade winds that brought Christopher Columbus to the New World. In the spring they would again head north, many stopping on the shores of Delaware Bay (also a link in the Western Hemisphere Shorebird Reserve Network) to fatten on horseshoe crab eggs.

The epic flights of these shorebirds have been knitting the hemispheres together, year after year, for time out of memory. They are a dramatic demonstration of the interconnectedness of the planet's ecosystems. This annual massing of birds stands as a heartening sign that despite the depredations of the last four centuries since Europeans first crossed the Atlantic, the natural abundance of this ocean can be preserved and renewed if care is taken to protect its critical habitats, north and south, along its magnificent coastline.

# PHOTOGRAPHER'S ACKNOWLEDGMENTS

I WOULD LIKE to acknowledge and extend my gratitude to the following-people and institutions who have assisted me in achieving the photography for this book: My wife and fellow photographer, Anne MacKay, for her love and support; Nancy Flight and the design staff of Greystone Books; Noel and Geri O'Dea for their kindness and hospitality; the Gatherall Family for their friendship and support in helping me capture many whale and seabird images; Richard Thomas, Manager of Mistaken Point Ecological Reserve; Parks and Natural Areas Division, Newfoundland; Danny Boyce, Ocean Sciences Centre, Memorial University of Newfoundland; Bill Robertson, Executive Director of The Huntsman Marine Science Centre, St. Andrews, New Brunswick; Bert Vissers, Shubenacadie Wildlife Park; Laurie Murison, Grand Manan Whale and Seabird Research Station; Chris Mansky and Sonja Wood of the Blue Beach Fossil Museum, Hansport, Nova Scotia. A contribution I greatly appreciate is from the Newfoundland and Labrador Department of Tourism. Over the years they have built a collection of my work, and it was kindly made available to me for this book. These images are reproduced on pages: 10, 27, 39–40, 48–49, 68, 107, 263, 264, 273, 276, and 292.

*Wayne Barrett*

# FURTHER READING

## ATLANTIC, GENERAL


Armstrong, Bruce. *Sable Island*. Toronto: Doubleday Canada, 1981.

Audubon, John James. *The Audubon Reader*. Edited by Richard Rhodes. New York: Alfred A. Knopf, 2006.

Berrill, Michael and Deborah Berrill. *The North Atlantic Coast: Cape Cod to Newfoundland*. San Francisco: Sierra Club Books, 1981.

Bertness, Mark D. *Atlantic Shorelines: Natural History and Ecology*. Princeton, NJ: Princeton University Press, 2007.

Bigelow, Henry Bryant and W.C. Schroeder. *Fishes of the Gulf of Maine*. Woods Hole, MA: Woods Hole Oceanographic Institution and Museum of Comparative Zoology, Harvard University, 1964.

Brown, Richard G.B. *Atlas of Eastern Canadian Seabirds*. Ottawa: Canadian Wildlife Service, 1986.

Cameron, Silver Donald. *The Living Beach*. Toronto: Macmillan Canada, 1998.

Cook, Ramsay, ed. *The Voyages of Jacques Cartier*. Toronto: University of Toronto Press, 1993.

DeGraaf, Richard M. and Mariko Yamasaki. *New England Wildlife: Habitat, Natural History, and Distribution*. Lebanon, NH: University Press of New England, 2001.

Denys, Nicolas. *The Description and Natural History of the Coasts of North America (Acadia)*. Translated and edited by William F. Ganong. Toronto: Champlain Society, 1908.

Gibson, Merritt. *Seashores of the Maritimes*. Halifax: Nimbus, 2003.

Gosner, Kenneth L. *A Field Guide to the Atlantic Seashore: Invertebrates and Seaweeds of the Atlantic Coast from the Bay of Fundy to Cape Hatteras*. Boston: Houghton Mifflin, 1978.

Hay, John and Peter Farb. *The Atlantic Shore: Human and Natural History from Long Island to Labrador*. New York: Harper & Row, 1966.

Katona, Steven K., Valerie Rough, and David T. Richardson. *A Field Guide to the Whales, Porpoises and Seals of the Gulf of Maine and Eastern Canada, Cape Cod to Newfoundland*. New York: Charles Scribner's Sons,1983.

MacLeish, William H. *The Gulf Stream: Encounters with the Blue God*. Boston: Houghton Mifflin, 1989.

McAlpine, Donald F. and Ian M. Smith, eds. *Assessment of Species Diversity in the Atlantic Maritime Ecozone*. Ottawa: NRC Research Press, 2010.

McAlpine, Donald F., Michael C. James, Jon Lien, and Stan A. Orchard. "Status and Conservation of Marine Turtles in Canadian Waters." *Herpetological Conservation* 2 (2007): 85–112.

McLaren, Ian A. and Christel Bell. *Birds of Sable Island*. Halifax: Nova Scotia Museum, 1972.
Neal, William J., Pilkey, Orrin H., and Kelley, Joseph T. *Atlantic Coast Beaches: A Guide to Ripples, Dunes, and Other Natural Features of the Seashore*. Missoula, MT: Mountain Press, 2007.
Odum, Eugene P. *Fundamentals of Ecology*. Philadelphia: W.B. Saunders, 1959.
Peterson, Roger Tory. *A Field Guide to the Birds East of the Rockies*. Boston: Houghton Mifflin, 1980.
Peterson, Roger Tory and James Fisher. *Wild America: The Record of a 30,000-Mile Journey around the Continent by a Distinguished Naturalist and His British Colleague*. Boston: Houghton Mifflin, 1955.
Pollock, Leland W. *A Practical Guide to the Marine Animals of Northeastern North America*. New Brunswick, NJ: Rutgers University Press, 1998.
Proctor, Noble S. and Patrick J. Lynch. *A Field Guide to North Atlantic Wildlife: Marine Mammals, Seabirds, Fish, and Other Sea Life*. New Haven, CT: Yale University Press, 2005.
Russell, Franklin. *The Secret Islands*. London: Hodder & Stoughton, 1965.
Saunders, Gary L. and Wayne Barrett. *Wildlife of Atlantic Canada and New England*. Halifax: Nimbus, 1991.
Schreiber, E.A. and Joanna Burger, eds. *Biology of Marine Birds*. Boca Raton, FL: CRC Press, 2001.
Shumway, Scott W. *Atlantic Seashore, Beach Ecology from the Gulf of Maine to Cape Hatteras*. Guilford, CT: Globe Pequot Press, 2008.
Slocum, Joshua. *Sailing Alone around the World and Voyage of the Liberdade*. New York: Macmillan, 1978.
Teal, John and Mildred Teal. *Life and Death of the Salt Marsh*. New York: Ballantine Books, 1969.
Tinbergen, Niko. *The Herring Gull's World: A Study of the Social Behavior of Birds*. New York: Lyons & Burford, 1989.
Tuck, James A. *Maritime Provinces Prehistory*. Ottawa: National Museum of Man, 1984.
Ulanski, Stan. *The Gulf Stream, Tiny Plankton, Giant Bluefin, and the Amazing Story of the Powerful River in the Atlantic*. Chapel Hill, NC: University of North Carolina Press, 2008.

#### GEOLOGY AND PALEONTOLOGY

Atlantic Geoscience Society. *The Last Billion Years: A Geological History of the Maritime Provinces of Canada*. Halifax: Nimbus, 2001.
Clapham, Matthew E., Guy M. Narbonne, and James G. Gehling. "Paleoecology of the Oldest Known Animal Communities: Ediacaran Assemblages at Mistaken Point, Newfoundland." *Paleobiology* 29, no. 4 (2003): 527–544.
Ferguson, Laing. *The Fossil Cliffs of Joggins*. Halifax: The Nova Scotia Museum, 1988.
Gould, Stephen Jay, general ed. *The Book of Life*. New York: Viking Penguin, 1993.
Oldale, Robert N. *Cape Cod, Martha's Vineyard and Nantucket: The Geologic Story*. Yarmouth Port, MA: On Cape Publications, 2001.
Olsen, P.E. "Discovery of Earliest Jurassic Reptile Assemblages from Nova Scotia Imply Catastrophic End to the Triassic." *Lamont Newsletter* 12 (1986).
Roberts, David C. *A Field Guide to Geology: Eastern North America*. Boston: Houghton Mifflin, 1996.
Sargeant, W.A.S. and Mossman, D.J. "Vertebrate Footprints from the Carboniferous Sediments of Nova Scotia: A Historical Review and Description of Newly Discovered Forms." *Paleogeography, Paleoclimatology, Paleoecology* 23 (1978): 279–306.

Thurston, Harry. *Dawning of the Dinosaurs: The Story of Canada's Oldest Dinosaurs*. Halifax: Nimbus and the Nova Scotia Museum, 1994.
Winchester, Simon. *Krakatoa: The Day the World Exploded: August 27, 1883*. New York: HarperCollins, 2003.

#### COASTAL FORESTS

Conner, William H., Thomas W. Doyle, and Ken W. Krauss, eds. *Ecology of Tidal Freshwater Forested Wetlands of the Southeastern United States*. Dordrecht, the Netherlands: Springer, 2007.
Fergus, Charles. *Trees of New England: A Natural History*. Guilford, CA: Globe Pequot Press, 2005.
Forman, Richard T.T. *Pine Barrens: Ecosystem and Landscape*. New York: Academic Press, 1979.
Henry, J. David. *Canada's Boreal Forest*. Washington, DC: Smithsonian Institution Press, 2002.
Laderman, Aimlee D., ed. *Coastally Restricted Forests*. New York: Oxford University Press, 1998.
Larsen, James A. *The Boreal Ecosystem*. New York: Academic Press, 1980.
Loo, Judy and N. Ives. "The Acadian forest: Historical condition and human impacts." *The Forestry Chronicle* 79, no. 3 (2003): 462–474.
McPhee, John. *The Pine Barrens*. New York: Farrar, Straus & Giroux, 1981.
Mosseler, A., J.A. Lynds, and J.E. Major. "Old-Growth Forests of the Acadian Forest Region." *Environmental Review* 11 (2003): S47–S77.
Ricketts, Taylor H., Eric Dinerstein, D. M. Olson, C.J. Loucks, et al., eds. *Terrestrial Ecoregions of North America: A Conservation Assessment*. Washington, DC: Island Press, 1999.
Yahner, Richard H. *Eastern Deciduous Forest: Ecology and Wildlife Conservation*. Minneapolis: University of Minnesota Press, 2000.

#### MID-ATLANTIC

Burger, Joanna. *A Naturalist along the New Jersey Shore*. New Brunswick, NJ: Rutgers University Press, 1996.
Curtin, Philip D., Grace S. Brush, and George W. Fisher, eds. *Discovering the Chesapeake: The History of an Ecosystem*. Baltimore: Johns Hopkins University Press, 2001.
Harrington, Brian. *The Flight of the Red Knot: A Natural History of a Small Bird's Annual Migration from the Arctic Circle to the Tip of South America and Back*. With Charles Flowers. New York: W.W. Norton, 1996.
Horton, Tom. *Turning The Tide: Saving the Chesapeake Bay*. Washington, DC: Island Press, 2003.
Leatherman, Stephen P. *Barrier Island Handbook*. Amherst, MA: University of Massachusetts, 1982.
Levinton, Jeffrey S. and John R. Waldman, eds. *The Hudson River Estuary*. New York: Cambridge University Press, 2006.
McPhee, John. *The Founding Fish*. New York: Farrar, Straus & Giroux, 2002.
Rappole, John H. *Wildlife of the Mid-Atlantic: A Complete Reference Manual*. Philadelphia: University of Pennsylvania Press, 2007.
Rountree, Helen C., Wayne E. Clark, and Kent Mountford, eds. *John Smith's Chesapeake Voyages, 1607–1609*. Charlottesville, VA: University of Virginia Press, 2007.
Sherman, Kenneth, N.A. Jaworski, T.J. Smayda, eds. *The Northeast Shelf Ecosystem: Assessment, Sustainability and Management*. Cambridge, MA: Blackwell Science, 1996.

Sutton, Clay and Pat Sutton. *Birds and Birding at Cape May*. Mechanicsburg, PA: Stackpole Books, 2006.
Warner, William W. *Beautiful Swimmers: Watermen, Crabs and the Chesapeake Bay*. Boston: Little, Brown, 1976.
Wells, John T. and Charles H. Peterson. *Restless Ribbons of Sand: Atlantic and Gulf Coast Barriers*. Baton Rouge, LA: Louisiana State University, 1986.

### GULF OF MAINE–BAY OF FUNDY

Conkling, Philip W., ed. *From Cape Cod to the Bay of Fundy: An Environmental Atlas of the Gulf of Maine*. Cambridge, MA: MIT Press, 1995.
Dadswell, M.J. and R.A. Rulifson. "Macrotidal Estuaries: A Region of Collision Between Migratory Marine Animals and Tidal Power Development." *Biological Journal of the Linnean Society* 51 (1994): 93–113.
Gordon, D.C., Jr., P.J. Cranford, and C. Desplanque. "Observations on the Ecological Importance of Salt Marshes in the Cumberland Basin, a Macrotidal Estuary in the Bay of Fundy." *Estuarine, Coastal and Shelf Science* 20 (1985): 205–207.
Hicklin, P.W. and P.C. Smith. "Selection of Foraging Sites and Invertebrate Prey by Migrant Semipalmated Sandpipers *Calidris pusilla (Pallas)* in Minas Basin, Bay of Fundy." *Canadian Journal of Zoology* 62 (1984): 2201–2210.
Hunstman, A.G. "The Production of Life in the Bay of Fundy." *Transactions of the Royal Society of Canada*. Series no. 3, section 5 (1952): 15–38.
Knowlton, A.R., S.D. Kraus, and R.D. Kenney. "Reproduction of North Atlantic Right Whales." *Canadian Journal of Zoology* 72 (1994): 1297–1305.
Peer, D.L., L.E. Linkletter, and P.W. Hicklin. "Life History and Reproductive Biology of *Corophium volutator* (Crustacea: Amphipoda) and the Influence of Shorebird Predation on Population Structure in Chignecto Bay, Bay of Fundy, Canada." *Netherlands Journal of Sea Research* 20, no. 4 (1986): 359–373.
Taylor, Peter H. *Salt Marshes in the Gulf of Maine: Human Impacts, Habitat Restoration, and Long-Term Change Analysis*. Gulf of Maine Council on the Marine Environment, 2008.
Thomas, Martin L.H., ed. *Marine and Coastal Systems of the Quoddy Region, New Brunswick*. Ottawa: Department of Fisheries and Oceans, 1983.
Thurston, Harry. *Tidal Life: A Natural History of the Bay of Fundy*. Camden East, ON: Camden House Publishing, 1990.
Tyrell, M.C. *Gulf of Maine Marine Habitat Primer*. Gulf of Maine Council on the Marine Environment, 2005.

### GULF OF ST. LAWRENCE

Canadian Wildlife Service. *Seabirds of Bonaventure Island*. Ottawa: Information Canada, 1973.
Dickie, L. M. and R.W. Trites. "The Gulf of St. Lawrence." In *Ecosystems of the World, No. 26: Estuaries and Enclosed Seas*, edited by B.H. Ketchum. New York: Elsevier Science, 1983.
El-Sabh, Mohammed I. and Norman Silverberg, eds. *Oceanography of a Large-Scale Estuarine System: The St. Lawrence*. New York: Springer-Verlag, 1990.
Gauthier, Jean and Jean Bédard. "Les déplacements de l'eider commun (*Somateria mollissima*) dans l'estuaire du Saint-Laurent." *Le Naturaliste canadien*, 103 (1976): 261–283.
Littlejohn, Bruce and Wayland Drew. *A Sea Within: The Gulf of St. Lawrence*. Toronto: McClelland and Stewart, 1984.

Michaud, R., A. Vegina, N. Rondeau, and Y. Vigneault. "Annual Distribution and Preliminary Characterization of Beluga Whale Habitats in the St. Lawrence." *Canadian Technical Report of Fisheries and Aquatic Sciences*, no. 1757 (1990).

Sergeant, D.E. *Harp Seals, Man and Ice*. Ottawa: Department of Fisheries and Oceans, 1991.

Smith, T.G., D.J. St. Aubin, and J.R. Geraci, eds. *Advances in Research on the Beluga Whale,* Delphinapterus leucas. Ottawa: Department of Fisheries and Oceans, 1990.

Therriault, Jean-Claude. *The Gulf of St. Lawrence: Small Ocean or Big Estuary?* Ottawa: Department of Fisheries and Oceans, 1991.

White, Louise and Frank Johns. *Marine Environmental Assessment of the Estuary and Gulf of St. Lawrence*. Ottawa: Department of Fisheries and Oceans, 1997.

## NEWFOUNDLAND AND LABRADOR

Brice-Bennett, Carol, ed. *Our Footprints Are Everywhere: Inuit Land Use and Occupancy in Labrador*. Nain, NL: Labrador Inuit Association, 1977.

Brown, Richard. *Voyage of the Iceberg: The Story of the Iceberg That Sank the Titanic*. Toronto: James Lorimer, 1983.

Jackson, Lawrence, ed. *Bounty of Barren Coast: Resource Harvest and Settlement in Southern Labrador*. St. John's, NL: Institute of Northern Studies, Memorial University for Petro Canada Explorations, 1982.

Leggett, W.C., K.T. Frank, and J.E. Carscadden. "Meteorological and Hydrographic Regulation of Year-Class Strength in Capelin (*Mallotus villosus*)." *Canadian Journal of Fisheries and Aquatic Sciences* 41, no. 81 (1984): 1193–1201.

MacPherson, Alan G. and Joyce B. MacPherson. *The Natural Environment of Newfoundland, Past and Present*. St. John's, NL: Memorial University of Newfoundland, 1981.

Nettleship, David N. and Tim R. Birkhead, eds. *The Atlantic Alcidae: Evolution, Distribution and Biology of the Auks Inhabiting the Atlantic Ocean and Adjacent Water Areas*. New York: Academic Press, 1985.

Robertson, G.J., S.I. Wilhelm, and P.A. Taylor. "Population Size and Trends of Seabirds Breeding on Gull and Great Islands, Witless Bay Islands Ecological Reserve, Newfoundland, up to 2003." *Canadian Wildlife Service Technical Report Series* (2004)

Rose, George A. *Cod: The Ecological History of the North Atlantic Fisheries*. St. John's, NL: Breakwater Books, 2007.

Templeman, Wilfred. *Life History of the Caplin (Mallotus villosus O.F. Muller) in Newfoundland Waters*. St. John's, NL: Newfoundland Government, 1948.

## CONSERVATION AND CLIMATE CHANGE

Baum, Julia K. and Boris Worm. "Cascading Top-Down Effects of Changing Oceanic Predator Abundances." *Journal of Animal Ecology* 18 (2009): 699–714.

Brown, M.W., J.A. Allen, and S.D. Kraus. "The Designation of Seasonal Right Whale Conservation Areas in the Waters of Atlantic Canada." In *Marine Protected Areas and Sustainable Fisheries*, edited by N.L Shackell and J.H.M. Willison. Wolfville, NS: Science & Management of Protected Areas Association, 1995.

Brown, R.G.B. "Marine birds and Climatic Warming in the Northwest Atlantic." In *Studies of High Latitude Seabirds*, edited by W.A. Montevecchi, A.J. Gaston, and R.D. Elliot. Ottawa: Canadian Wildlife Service, 1991.

Casey, Jill M. and Ransom A. Myers. "Near Extinction of a Large, Widely Distributed Fish." *Science* 281 (1998): 690–692.

Hitchings, Jeffrey A. and Ransom A. Myers. "What Can Be Learned from the Collapse of a Renewable Resource? Atlantic Cod, *Gadus morhua*, of Newfoundland and Labrador." *Canadian Journal Fisheries Aquatic Sciences* 51 (1994): 2126–2146.

Lotze, Heike K. "Rise and Fall of Fishing and Marine Resource Use in the Wadden Sea, Southern North Sea." *Fisheries Research* 87 (2007): 208–218.

Lotze, Heike K. and Inka Milewski. "Two Centuries of Multiple Impacts and Successive Changes in a North Atlantic Food Web." *Ecological Applications* 14, no. 5 (2004): 1428–1447.

Mitchell, Alanna. *Seasick: Ocean Change and the Extinction of Life on Earth*. Chicago: University of Chicago Press, 2009.

Mowat, Farley. *Sea of Slaughter*. Toronto: Key Porter Books, 2003.

Myers, Ransom A., J.K. Baum, T.D. Shepherd, S.P. Powers, and C.H. Peterson. "Cascading Effects of the Loss of Apex Predatory Sharks from a Coastal Ocean." *Science* 315 (2007): 1846–1850.

Myers, Ransom A., and B. Worm. "Extinction, Survival, or Recovery of Large Predatory Fishes." *Proceedings of the Royal Society* 360 (2005): 13–20.

Myers, Ransom A., and B. Worm. "Rapid Worldwide Depletion of Predatory Fish Communities." *Nature* 423 (2003): 280–283.

Nye, Janet. *Climate Change and Its Effects on Ecosystems, Habitats and Biota: State of the Gulf of Maine Report*. Gulf of Maine Council on the Marine Environment, June 2010.

Pauly, Daniel, V. Christensen, J. Dalsgaard, R. Froese, and F. Torres, Jr. "Fishing Down Marine Food Webs." *Science* 279 (1998): 860–863.

Whitehead, Hal, B. McGill, and B. Worm. "Diversity of Deep-Water Cetaceans in Relation to Temperature: Implications for Ocean Warming." *Ecology Letters* 11 (2008): 1198–1207.

Worm, B., R. Hilborn, J.K. Baum, T.A. Branch, J.S. Collie, C. Costello, M.J. Fogarty, et al. "Rebuilding Global Fisheries." *Science* 325 (2009): 578–585.

Worm, B., E.B. Barbier, N. Beaumont, J.E. Duffy, C. Folke, B.S. Halpern, J.B.C. Jackson, et al. "Impacts of Biodiversity Loss on Ocean Ecosystem Services." *Science* 314 (2006): 787–790.

Worm, B. and Ransom A. Myers. "Meta-analysis of Cod-Shrimp Interactions Reveals Top-Down Control in Oceanic Food Webs." *Ecology* 84 (2003): 162–173.

# SCIENTIFIC NAMES

Common names for plants and animals are given in the book without their scientific names, which are provided here. These entries are listed alphabetically by common name; for example, pitch pine will be under P and red pine under R.

alder flycatcher, *Empidonax alnoram*
alewife or gaspereau, *Alosa pseudoharengus*
American beach grass or marram grass, *Ammophila breviligulata*
American beech, *Fagus grandifolia*
American bittern, *Botaurus lentiginosus*
American black duck, *Anas rubripes*
American bulrush, *Scirpus pungens*
American butterfish, *Peprilus triacanthus*
American chestnut, *Castanea dentata*
American conger, *Conger oceanica*
American eel, *Anguilla rostrata*
American holly, *Ilex opaca*
American hornbeam, *Carpinus caroliniana*
American lobster, *Homarus americanus*
American mountain ash, *Sorbus americana*
American oystercatcher, *Haematopus palliatus*
American plaice or flounder, *Hippoglossoides platessoides*
American redstart, *Setophaga ruticilla*
American robin, *Turdus migratorius*
American sand lance, *Ammodytes americanus*
American shad, *Alosa sapidissima*
American smelt, *Osmerus mordax*
American widgeon, *Anas americana*
Arctic char, *Salvelinus alpines*
Arctic hare, *Lepus arcticus*
Arctic tern, *Sterna paradisaea*
argentine. *See* Atlantic argentine
arrow arum, *Peltandra virginica*
Asian spiderwort, *Murdannia keisak*
Atlantic argentine, *Argentina silus*
Atlantic cod, *Gadus morhua*
Atlantic halibut, *Hippoglossus hippoglossus*
Atlantic herring, *Clupea harengus*
Atlantic mackerel, *Scomber scombrus*
Atlantic menhaden, *Brevoortia tyrannus*
Atlantic puffin, *Fratercula arctica*
Atlantic salmon, *Salmo salar*
Atlantic silverside, *Menidia menidia*
Atlantic sturgeon, *Acipenser oxyrhynchus*
Atlantic surf clam, *Spisula solidissima*
Atlantic tomcod, *Microgadus tomcod*
Atlantic walrus, *Odobenus rosmarus rosmarus*
Atlantic white cedar, *Chamaecyparis thyoides*
Atlantic white-sided dolphin, *Lagenorhynchus acutus*
Audubon's shearwater, *Puffinus lherminieri*
bald cypress, *Taxodium distichum*
bald eagle, *Haliaeetus leucocephalus*
balsam fir, *Abies balsamea*
balsam poplar, *Populus balsamifera*
barndoor skate, *Dipturus laevis*
barred owl, *Strix varia*
bay anchovy, *Anchoa mitchilli*
bay-breasted warbler, *Dendroica castanea*
bay scallop, *Argopecten irradians*
beach heather, *Hudsonia tomentosa*
beach plum, *Prunus maritima*
bearded seal, *Erignathus barbatus*
beaver, *Castor canadensis*

beluga, *Delphinapterus leucas*
Bicknell's thrush, *Catharus bicknelli*
big bluestem, *Andropogon gerardii*
bitternut hickory, *Carya cordiformis*
black-and-white warbler, *Mniotilta varia*
black bear, *Ursus americanus*
black-bellied plover, *Pluvialis squatarola*
Blackburnian warbler, *Dendroica fusca*
black-capped chickadee, *Parus atricapillus*
black cherry, *Prunus serotina*
black drum, *Pogonias cromis*
black duck, *Anas rubripes*
black guillemot, *Cepphus grylle*
black-legged kittiwake, *Rissa tridactyla*
blackpoll warbler, *Dendroica striata*
black scoter or common scoter, *Melanitta nigra*
black sea bass, *Centropristis striata*
black skimmer, *Rynchops niger*
black spruce, *Picea mariana*
black-throated blue warbler, *Dendroica caerulescens*
black-throated green warbler, *Dendroica virens*
black tupelo, *Nyssa sylvatica*
bladder wrack, *Fucus vesiculosus*
blue crab, *Callinectes sapidus*
blue green algae, *Calothrix* spp.
blue jay, *Cyanocitta cristata*
blue mussel, *Mytilus edulis*
blueback herring, *Alosa aestivalis*
bluefin tuna, *Thunnus thynnus*
bluefish, *Pomatomus saltatrix*
blue-gray gnatcatcher, *Polioptila caerulea*
blue whale, *Balaenoptera musculus*
blue-winged teal, *Anas discors*
bobcat, *Lynx rufus*
bobolink, *Dolichonyx oryzivorus*
Bonaparte's gull, *Larus philadelphia*
boreal owl, *Aegolis funereus*
bottlenose dolphin, *Tursiops truncatus*
bowhead whale, *Balaena mysticetus*
brant, *Branta bernicla*
British storm petrel, *Hydrobates pelagicus*
brook trout, *Salvelinus fontinalis*
broom sedge, *Andropogon virginicus*
brown thrasher, *Toxostoma rufum*
bufflehead, *Bucephala albeola*
bull shark, *Carcharhinus leucas*
bur oak, *Quercus macrocarpa*
Canada goose, *Branta canadensis*
Cape May warbler, *Dendroica tigrina*
capelin, *Mallotus villosus*
caribou, *Rangifer tarandus*
Carolina chickadee, *Poecile carolinensis*
Carolina wren, *Thryothorus ludovicianus*
Caspian tern, *Sterna caspia*
cattle egret, *Bubulcus ibis*
chestnut blight fungus, *Cryphonectria parasitica*
chestnut oak, *Quercus prinus*
chestnut-sided warbler, *Dendroica pensylvanica*
common cockle, *Cerastoderma edule*
common eider, *Somateria mollissima*
common goldeneye, *Bucephala clangula*
common hair grass, *Deschampsia flexuosa*
common loon, *Gavia immer*
common merganser, *Mergus merganser*
common murre, *Uria aalge*
common nighthawk, *Chordeiles minor*
common periwinkle, *Littorina littorea*
common raven, *Corvus corax*
common scoter. *See* black scoter
common tern, *Sterna hirundo*
common wild rose, *Rosa virginiana*
conger eel. *See* American conger
copepod, *Calanus finmarchicus*
corn snake, *Elaphe guttata*
Cory's shearwater, *Puffinus diomedea*
cottontail rabbit. *See* eastern cottontail
cougar, *Puma concolor*
cownose ray, *Rhinoptera bonasus*
coyote, *Canis latrans*
crabeater seal, *Lobodon carcinophagus*
crowberry, *Empetrum nigrum*
cunner, *Tautogolabrus adspersus*
cusk, *Brosme brosme*
dead man's fingers, *Alcyonium digitatum*
diamondback terrapin, *Malaclemys terrapin*
dogtooth violet, *Erythronium dens-canis*
dog whelk, *Nucella lapillus*
double-crested cormorant, *Phalacrocorax auritus*
dovekie, *Alle alle*
downy woodpecker, *Picoides pubescens*
drum. *See* black drum, red drum
dunlin, *Calidris alpina*
eastern chipmunk, *Tamias striatus*
eastern cottontail, *Sylvilagus floridanus*
eastern cougar. *See* cougar
eastern coyote. *See* coyote
eastern harlequin duck. *See* harlequin duck
eastern hemlock, *Tsuga canadensis*

eastern meadowlark, *Sturnella magna*
eastern oyster, *Crassostrea virginica*
eastern towhee. *See* rufous-sided towhee
eastern white cedar, *Thuja occidentalis*
eastern white pine, *Pinus strobus*
eastern wood-pewee, *Contopus virens*
eelgrass, *Zostera marina*
elk, *Cervus canadensis*
European green crab, *Carcinus maenas*
evening grosbeak, *Hesperiphona vespertina*
fiddler crab, *Uca* sp
fin whale, *Balaenoptera physalus*
fire cherry, *Prunus pensylvanica*
fisher, *Martes pennanti*
flounder. *See* American plaice
Forster's tern, *Sterna forsteri*
fox sparrow, *Passerella iliaca*
gadwall, *Anas strepera*
giant squid, *Architeuthis dux*
glasswort, *Salicornia* sp
glossy ibis, *Plegadis falcinellus*
golden crowned kinglet, *Regulus satrapa*
goosefoot, *Atriplex patula*
grasshopper sparrow, *Ammodramus savannarum*
gray phalarope, *Phalaropus fulicarius*
gray seal, *Halichoerus grypus*
gray squirrel, *Sciurus carolinensis*
gray wolf, *Canus lupus*
great auk (extinct), *Pinguinus impennis*
great black-backed gull, *Larus marinus*
great blue heron, *Ardea herodias*
great cormorant, *Phalacrocorax carbo*
great egret, *Casmerodius albus*
greater scaup, *Aythya marila*
greater shearwater, *Puffinus gravis*
greater snow goose, *Chen caerulescens*
greater yellowlegs, *Tringa melanoleuca*
great horned owl, *Bubo virginianus*
great white shark, *Carcharodon carcharias*
green ash, *Fraxinus pennsylvannica*
green sea urchin, *Strongylocentrotus droebachiensis*
green turtle, *Chelonia mydas*
Greenland halibut, *Reinhardtius hippoglossoides*
green-winged teal, *Anas carolinensis*
gull-billed tern, *Gelochelidon nilotica*
haddock, *Melanogrammus aeglefinus*
halibut. *See* Atlantic halibut, Greenland halibut
harbor porpoise, *Phocoena phocoena*
harbor seal, *Phoca vitulina*
harlequin duck, *Histrionicus histrionicus*
harp seal, *Phoca groenlandica*
herring gull, *Larus argentatus*
hickory shad, *Alossa mediocris*
hooded seal, *Crystophora cristata*
hooded warbler, *Wilsonia citrina*
horned lark, *Eremophila alpestris*
horseshoe crab, *Limulus polyphemus*
humpback whale, *Megaptera novaeangliae*
Indian grass, *Sorghastrum nutans*
Ipswich sparrow, *Passerculus sandwichensis princeps*
ivory gull, *Pagophila eburnea*
jack pine, *Pinus banksiana*
junco, *Junco* spp.
kelp, *Laminaria* sp.
Kemp's ridley turtle, *Lepidochelys kempii*
killer whale, *Orcinus orca*
king eider, *Somateria spectabilis*
kinglet. *See* golden-crowned kinglet, ruby-crowned kinglet
knotted wrack, *Ascophyllum nodosum*
krill, *Thysanoessa raschii* and *Meganyctiphanes norvegica*
least tern, *Sternula antillarum*
Labrador duck (extinct), *Camptorhyncus labradorius*
Labrador tea, *Ledum latifolium*
laughing gull, *Larus atricilla*
laurel, *Kalmia glauca*
Leach's storm petrel, *Oceanodroma leucorhoa*
least flycatcher, *Empidonax minimus*
little blue heron, *Florida caerula*
little bluestem, *Schizachyrium scoparium*
loggerhead turtle, *Caretta caretta*
longfin squid, *Loligo pealeii*
longhorn sculpin, *Myoxocephalus octodecemspinosus*
Long's Bullrush, *Scirpus longii*
long-tailed duck, *Clangula hyemalis*
lumpfish, *Cyclopterus lumpus*
lung worm, *Sternurus* sp
lynx, *Lynx canadensis*
Manx shearwater, *Puffinus puffinus*
Maritime ringlet, *Coenonympha nipisiquit*
marram grass. *See* American beach grass
marsh periwinkle, *Littorina irrorata*
marten, *Martes americana*
meadow jumping mouse, *Zapus hudsonius*
menhaden. *See* Atlantic menhaden
merlin, *Falco columbarius*
mink, *Mustela vison*
minke whale, *Balaenoptera acutorostrata*

mockernut hickory, *Carya tomentosa*
monkfish, *Lophius americanus*
moon snail. *See* northern moon snail
moose, *Alces alces*
mountain maple, *Acer spiatum*
mud shrimp, *Corophium volutator*
muskrat, *Ondatra zibethicus*
naked goby, *Gobiosoma bosc*
New Jersey rush, *Juncus caesariensis*
northern bog lemming, *Synaptomys borealis*
northern bottlenose whale, *Hyperoodon ampullatus*
northern flying squirrel, *Glaucomys sabrinus*
northern fulmar, *Fulmarus glacialis*
northern gannet, *Morus bassanus*
northern moon snail, *Lunatia heros*
northern phalarope, *Lobipes lobatus*
northern pine snake, *Pituophis melanoleucus*
northern pintail, *Anas acuta*
northern red oak, *Quercus rubra*
northern right whale, *Eubalaena glacialis*
northern sand lance, *Ammodytes dubius*
northern shoveler, *Anas clypeata*
nutria, *Myocastor coypus*
olive-sided flycatcher, *Nuttallornis borealis*
orchard oriole, *Icterus spurius*
osprey, *Pandion haliaetus*
ovenbird, *Seiurus aurocapillus*
oyster toad fish, *Opsanus tau*
palm warbler, *Dendroica palmarum*
pandalid shrimp, *Pandalus* spp.
paper birch. *See* white birch
peregrine falcon, *Falco peregrinus*
pied-billed grebe, *Podilymbus podiceps*
pignut hickory, *Carya glabra*
pileated woodpecker, *Dryocopus pileatus*
pilot whale, *Globicephala melaena*
Pine Barrens treefrog, *Hyla andersonii*
pine grosbeak, *Pinicola enucleator*
pine siskin, *Carduelis pinus*
pine snake. *See* northern pine snake
pink coreopsis, *Coreopsis rosea*
pink lady's slipper, *Cypripedium acaule*
pintail, *Anas acuta*
piping plover, *Charadrius melodus*
pitch pine, *Pinus rigida*
Plymouth gentian, *Sabatia kennedyana*
pollack, *Pollachius virens*
pond pine, *Pinus serotina*
porbeagle shark, *Lamna nasus*
porcupine, *Erethizon dorsatum*
poverty grass. *See* beach heather
prairie warbler, *Dendroica discolor*
prothonotary warbler, *Protonotaria citrea*
pygmy blue whale, *Balaenoptera musculus brevicauda*
raccoon, *Procyon lotor*
rainbow smelt, *Osmerus mordax*
raven. *See* common raven
razorbill, *Alca torda*
red-backed vole. *See* southern red-backed vole
red bay, *Persea borbonia*
red-bellied woodpecker, *Melanerpes carolinus*
red-breasted merganser, *Mergus serrator*
red crossbill, *Loxia curvirostra*
red drum, *Sciaenops ocellatus*
red-eyed vireo, *Vireo olivaceus*
redfish, *Sebastes marinus*
red fox, *Vulpes fulva*
red hake, *Urophycis chuss*
redhead duck, *Aythya americana*
redhead grass, *Potamogeton perfoliatus*
red knot, *Calidris canutus*
red maple, *Acer rubrum*
red-necked phalarope, *Phalaropus lobatus*
red pine, *Pinus resinosa*
red spruce, *Picea rubens*
red squirrel, *Tamiasciurus hudsonicus*
red-throated loon, *Gavia stellata*
red-winged blackbird, *Agelaius phoeniceus*
reindeer lichen, *Cladonia rangiferina*
ribbed mussel, *Geukensia demissa*
ringed seal, *Pusa hispida*
river otter, *Lutra canadensis*
rock crab, *Cancer irroratus*
roseate tern, *Sterna dougallii*
rough periwinkle, *Littorina saxatilis*
royal tern, *Sterna maxima*
ruby-crowned kinglet, *Regulus calendula*
ruddy duck, *Oxyura jamaicensis*
ruddy turnstone, *Arenaria interpres*
ruffed grouse, *Bonasa umbellus*
rufous-sided towhee, *Pipilo erythrophthalmus*
rusty blackbird, *Euphagus carolinus*
sago pondweed, *Stuckenia pectinata*
salt marsh bulrush, *Scirpus robustus*
salt marsh hay, *Spartina patens*
salt marsh snail, *Melampus bidentatus*
sand lance. *See* American sand lance, northern sand lance
sand shrimp, *Crangon septemspinosa*

sandbar shark, *Carcharhinus plumbeus*
sanderling, *Calidris alba*
Savannah sparrow, *Passerculus sandwichensis*
saxifrage, *Saxifraga* spp.
scarlet tanager, *Piranga olivacea*
scrub oak, *Quercus ilicifolia*
scud. *See* sideswimmer
sea horse, *Hippocampus* spp.
sea lamprey, *Petromyzon marinus*
sea lettuce, *Ulva lactuca*
sea mink (extinct), *Neovison macrodon*
seaside goldenrod, *Solidago sempervirens*
seaside plantain, *Plantago maritima*
sei whale, *Balaenoptera borealis*
semipalmated plover, *Charadrius semipalmatus*
semipalmated sandpiper, *Calidris pusilla*
shagbark hickory, *Carya ovata*
short-beaked common dolphin, *Delphinus delphis*
short-billed dowitcher, *Limnodromus scolopaceus*
short-eared owl, *Asio flammeus*
shortnose sturgeon, *Acipenser brevirostrum*
sideswimmer or scud, *Gammarus oceanicus*
silver hake, *Merluccius bilinearis*
silverweed, *Argentina anserina*
skilletfish, *Gobiesox strumosus*
slime mold, *Labyrinthula* spp.
small yellow lady's slipper, *Cypripedium parviflorum*
smooth cordgrass, *Spartina alterniflora*
snowshoe hare, *Lepus americanus*
snowy egret, *Egretta thula*
sooty shearwater, *Puffinus griseus*
southern red-backed vole, *Clethrionomys gapperi*
southern red oak, *Quercus falcata*
sperm whale, *Physeter catodon*
sphagnum moss, *Sphagnum* spp.
spiny dogfish, *Squalus acanthias*
spring beauty, *Claytonia virginica*
spring peeper, *Hyla crucifer*
spruce grouse, *Canachites canadensis*
striped bass, *Morone saxatilis*
striped blenny, *Meiacanthus grammistes*
striped dolphin, *Stenella coeruleoalba*
sugar maple, *Acer saccharum*
summer flounder, *Paralichthys dentatus*
surf clam. *See* Atlantic surf calm
surf scoter, *Melanitta perspicillata*
Swainson's thrush, *Catharus ustulatus*
swamp tupelo. *See* black tupelo
sweetbay, *Magnolia virginiana*
switchgrass, *Panicum virgatum*
swordfish, *Xiphias gladius*
tamarack, *Larix laricina*
Tennessee warbler, *Vermivora peregrina*
thick-billed murre, *Uria lomvia*
thorny skate, *Amblyraja radiata*
tiger shark, *Galeocerdo cuvier*
timber rattlesnake, *Crotalus horridus*
trembling aspen, *Populus tremuloides*
tundra swan, *Cygnus columbianus*
turkey vulture, *Cathartes aura*
upland sandpiper, *Bartramia longicauda*
Virginia creeper, *Parthenocissus quinquefolia*
walrus. *See* Atlantic walrus
white birch, *Betula papyrifera*
white-breasted nuthatch, *Sitta carolinensis*
white-footed mouse, *Peromyscus leucopus*
white hake, *Urophycis tenuis*
white mullet, *Mugil curema*
white oak, *Quercus alba*
white perch, *Morone americana*
white spruce, *Picea glauca*
white-tailed deer, *Odocoileus virginianus*
white-winged crossbill, *Loxia leucoptera*
white-winged scoter, *Melanitta deglandi*
widgeon grass, *Ruppia maritime*
wild celery, *Vallisneria americana*
wild turkey, *Meleagris gallopavo*
willet, *Catoptrophorus semipalmatus*
willow oak, *Quercus phellos*
Wilson's storm petrel, *Oceanites oceanicus*
winter flounder, *Pseudopleuronectes americanus*
winter wren, *Troglodytes hiemalis*
wolverine, *Gulo gulo*
woodchuck, *Marmota monax*
woodland vole, *Microtus pinetorum*
wood thrush, *Hylocichla mustelina*
wood warbler, *Phylloscopus sibiliatrix*
worm-eating warbler, *Helmitheros vermivorus*
yellow birch, *Betula lutea* in Canada, *Betula alleghaniensis* in the United States
yellow-breasted chat, *Icteria virens*
yellow lady's slipper. *See* small yellow lady's slipper
yellow-rumped warbler, *Dendroica coronata*
yellowtail flounder, *Limanda ferruginea*
yellow warbler, *Dendroica petechia*

# INDEX

Page numbers in bold indicate photographs, illustrations, or tables. Species are listed by their common name unless only the Latin name appears in the text. Refer to the Scientific Names list for a complete listing of common names and Latin names.

**THE DAVID SUZUKI FOUNDATION**

The David Suzuki Foundation works through science and education to protect the diversity of nature and our quality of life, now and for the future.

With a goal of achieving sustainability within a generation, the Foundation collaborates with scientists, business and industry, academia, government and non-governmental organizations. We seek the best research to provide innovative solutions that will help build a clean, competitive economy that does not threaten the natural services that support all life.

The Foundation is a federally registered independent charity that is supported with the help of over 50,000 individual donors across Canada and around the world.

We invite you to become a member. For more information on how you can support our work, please contact us:

The David Suzuki Foundation
219–2211 West 4th Avenue
Vancouver, BC
Canada V6K 4S2
www.davidsuzuki.org
contact@davidsuzuki.org
Tel: 604-732-4228
Fax: 604-732-0752

Checks can be made payable to The David Suzuki Foundation.
All donations are tax-deductible.

Canadian charitable registration: (BN) 12775 6716 RR0001
U.S. charitable registration: #94-3204049